International Code
of Phylogenetic
Nomenclature
(PhyloCode)

International Code of Phylogenetic Nomenclature (PhyloCode)

Version 6*

Philip D. Cantino and Kevin de Queiroz**

Ratified on January 20, 2019, by the Committee on Phylogenetic Nomenclature of the International Society for Phylogenetic Nomenclature: Sina M. Adl, Philip D. Cantino, Nico Cellinese, Kevin de Queiroz, James A. Doyle, Micah Dunthorn, Sean W. Graham, Max Cardoso Langer, Michel Laurin, Richard G. Olmstead, George Sangster, and Mieczyslaw Wolsan.

* This is the first version published in hard copy. Previous versions are available at www.phylocode.org.
** Authors listed alphabetically

CRC Press
Taylor & Francis Group
Boca Raton London New York

CRC Press is an imprint of the
Taylor & Francis Group, an **informa** business

CRC Press
Taylor & Francis Group
6000 Broken Sound Parkway NW, Suite 300
Boca Raton, FL 33487-2742

© 2020 by Taylor & Francis Group, LLC
CRC Press is an imprint of Taylor & Francis Group, an Informa business

No claim to original U.S. Government works

First published in digital version April 29, 2020.

International Standard Book Number-13: 978-1-138-33282-9 (Paperback)
International Standard Book Number-13: 978-1-138-33286-7 (Hardback)

Library of Congress Cataloging-in-Publication Data

Names: Cantino, Philip D., author. | De Queiroz, Kevin, author. |
International Society for Phylogenetic Nomenclature. Committee on
Phylogenetic Nomenclature.
Title: International code of phylogenetic nomenclature : PhyloCode / Philip
D. Cantino and Kevin de Queiroz.
Other titles: PhyloCode
Description: Version 6. | Boca Raton : CRC Press, 2020. | Ratified on
January 20, 2019, by the Committee on Phylogenetic Nomenclature, of the
International Society for Phylogenetic Nomenclature | Includes
bibliographical references. | Summary: "This book will govern the names
of clades, while species names will still be governed by traditional
codes. PhyloCode is designed so that it can be used concurrently with
the rankbased codes. It is not meant not to replace existing systems but
to provide an alternative system for governing the application of both
existing and newly proposed names"-- Provided by publisher.
Identifiers: LCCN 2019028676 (print) | LCCN 2019028677 (ebook) | ISBN
9781138332829 (paperback) | ISBN 9781138332867 (hardback) | ISBN
9780429446320 (ebook)
Subjects: LCSH: Biology--Classification. | Biology--Nomenclature. |
Cladistic analysis. | Phylogeny--Nomenclature. |
Plants--Phylogeny--Nomenclature.
Classification: LCC QH83 .C28 2020 (print) | LCC QH83 (ebook) | DDC
570.1/2--dc23
LC record available at https://lccn.loc.gov/2019028676
LC ebook record available at https://lccn.loc.gov/2019028677

Visit the Taylor & Francis Web site at
http://www.taylorandfrancis.com

and the CRC Press Web site at
http://www.crcpress.com

Contents

CONTENTS

Preface

Version 6 is the first version published as a printed volume. Previous versions were solely electronic and are available at www.phylocode. org. The material in this Preface has been summarized from a variety of sources; see the History section for literature citations.

The development of the *International Code of Phylogenetic Nomenclature* (referred to here as the *PhyloCode*) grew out of the recognition that the current rank-based systems of nomenclature, as embodied in the current botanical, zoological, and bacteriological codes, are not well suited to govern the names of clades. Clades (along with species) are the entities that make up the tree of life, and for this reason they are among the most theoretically significant biological entities above the organism level. Consequently, clear communication and efficient storage and retrieval of biological information require names that explicitly and unambiguously refer to clades and do not change over time. The current rank-based codes fail to provide such names for clades. Supraspecific names are not always associated with clades under the rank-based codes, and even when they are, they often fail to retain their associations with particular clades because the names are implicitly defined in terms of ranks and types. A clade whose hypothesized composition and diagnostic characters have not changed may be given a different name under the rank-based codes based purely on considerations of rank. Such instability is particularly objectionable given the wide recognition that rank assignment is subjective and of dubious biological significance.

In contrast to the rank-based codes, the *PhyloCode* provides rules for the express purpose of naming clades through explicit reference to phylogeny. In doing so, the *PhyloCode* extends "tree-thinking" to biological nomenclature. This development parallels the extension of tree-thinking into taxonomy, as manifested in the concepts of species as lineage segments and supraspecific taxa as clades. These nomenclatural and taxonomic developments are complementary but independent. Clades can be named using the traditional rank-based systems of nomenclature (though with the problems noted above), and a nomenclatural system based on phylogenetic principles does not require equating supraspecific taxa with clades. The *PhyloCode*, however, is designed for the specific purpose of naming clades.

The objective of the *PhyloCode* is not to replace existing names but to provide an alternative system for governing the application of both existing and newly proposed names. In developing the *PhyloCode*, much thought has been given to minimizing disruption of the existing nomenclature. Thus, rules and recommendations have been included to ensure that most names will be applied in ways that approximate their current and/or historical use. However, names that apply to clades will be redefined in terms of phylogenetic relationships rather than taxonomic rank and therefore will not be subject to the subsequent changes that occur under the rank-based systems due to changes in rank. Because the taxon membership associated with particular names will sometimes differ between rank-based and phylogenetic systems, suggestions are provided for indicating which code governs a name when there is a possibility of confusion. Mechanisms are also provided to reduce certain types of nomenclatural divergence relative to the rank-based systems. For example, if a clade name is based on a genus name, the type of the genus under the appropriate rank-based code must be used as an internal specifier under the *PhyloCode* (Article 11.10, Examples 1 and 2).

The starting date of the *PhyloCode* coincides with the publication of *Phylonyms*, a volume that provides phylogenetic definitions for many widely used clade names and the names of many large clades (see below). Names for which phylogenetic definitions were published before that date, and not subsequently, are not considered established under this code.

Properties of Phylogenetic Nomenclature. The phylogenetic system of nomenclature embodied in the *PhyloCode* exhibits both similarities to and differences from the rank-based systems embodied in the traditional codes. Some of the most important similarities are as follows: (1) Both systems have the same fundamental goals of providing unambiguous methods for applying names to taxa, selecting a single accepted name for a taxon from among competing synonyms or homonyms, and promoting nomenclatural stability and continuity to the extent that doing so does not contradict new results and conclusions. (2) Neither system infringes upon the judgment of taxonomists with respect to inferring the composition of taxa or to assigning taxonomic ranks. (3) Both systems use precedence, a clear order of preference, to determine the correct name of a taxon when synonyms or homonyms exist. (4) Both systems use the date of publication (chronological priority) as the primary criterion for establishing precedence. (5) And both phylogenetic and rank-based systems have conservation mechanisms that allow a later-established name to be given precedence over an earlier name for the same taxon if using the earlier name would be contrary to the fundamental goal of promoting nomenclatural stability and continuity.

Some of the most important differences between the phylogenetic system of the *PhyloCode* and the rank-based systems of the traditional codes are as follows: (1) The phylogenetic system is independent of taxonomic rank. Although clades form nested hierarchies, the assignment of taxonomic rank is not part of the naming process and has no bearing on the spelling or application of clade names. As a consequence, the phylogenetic system does not require ranked taxonomies. (2) All taxa named under this code are clades. Clades are products of evolution that have an objective existence regardless of whether they are named. As a consequence, once a clade is named and its name associated with a phylogenetic definition, its composition and diagnostic characters become questions to be decided by empirical evidence rather than by personal decisions. (3) In addition to applying names to nested and mutually exclusive taxa, as in traditional nomenclature, the phylogenetic system allows names to be applied to partially overlapping clades. This provision is necessary to accommodate situations involving clades of hybrid origin. (4) In contrast to the rank-based codes, which use (implicit) definitions based on ranks and types to determine the application of names, phylogenetic nomenclature uses explicit phylogenetic definitions. Species, specimens, and apomorphies cited within these definitions are called *specifiers* because they are used to specify the clade to which the name applies. These specifiers function analogously to the types of rank-based nomenclature in providing reference points that determine the application of a name; however, they differ from types in that they may either be included in or excluded from the taxon being named, and multiple specifiers may be used. (5) The fundamental difference between the phylogenetic and rank-based systems in how names are defined leads to differences in how synonyms and homonyms are determined in practice. For example, under the *PhyloCode*, synonyms are names whose phylogenetic definitions specify the same clade, regardless of prior associations with particular ranks; in contrast, under the rank-based codes, synonyms are names at the same rank whose types are included within a single taxon at that rank, regardless of prior associations with particular clades. (6) Another novel aspect of the *PhyloCode* is that it permits taxonomists to restrict the application of names with respect to clade composition. If a taxonomist wishes to ensure that a name refers to a clade that either includes or excludes particular taxa, this result may be achieved through the use of additional internal or external specifiers (beyond the minimal number needed to specify a clade), or the definition may contain a qualifying clause specifying conditions under which the name cannot be used. (7) The *PhyloCode* includes recommended naming conventions that promote an integrated system of names for crown and total clades. The resulting pairs of names (e.g., *Testudines* and *Pan-Testudines* for the turtle crown and total clades, respectively) enhance the cognitive efficiency of the system and provide hierarchical information. (8) Establishment of a name under the *PhyloCode* requires both publication and registration. The purpose of registration is to create a comprehensive database of established names (discussed below), which will reduce the frequency of accidental homonyms and facilitate the retrieval of nomenclatural information.

Advantages of Phylogenetic Nomenclature. Phylogenetic nomenclature has several advantages over the traditional systems. It eliminates a major source of instability under the rank-based codes—changes in clade names due solely to shifts in rank. It also facilitates the naming of new clades as they are discovered. Under the rank-based codes, it is often difficult to name clades one at a time, similar to the way that species are named, because the name of a taxon is affected by the taxon's rank, which in turn depends on the ranks of more and less inclusive taxa. In a group in which the standard ranks are already in use, naming a newly discovered clade requires either the use of an unconventional intermediate rank (e.g., supersubfamily) or the shifting of less or more inclusive clades to lower or higher ranks, thus causing a cascade of name changes. This situation discourages systematists from naming clades until an entire classification is developed. In the meanwhile, well-supported clades are left unnamed, and taxonomy falls progressively farther behind knowledge of phylogeny. This is a particularly serious drawback at the present time, when advances in molecular and computational biology have led to a burst of new information about phylogeny, much of which is not being incorporated into taxonomy. The availability of the *PhyloCode* will permit researchers to name newly discovered clades much more easily than they can under the rank-based codes. For many researchers, naming clades is just as important as naming species. In this respect, the *PhyloCode* reflects a philosophical shift from naming species and subsequently classifying them (i.e., into higher taxa) to naming both species and clades. This does not mean, however, that all clades must be named. The decision to name a clade (or to link an existing name to it by publishing a phylogenetic definition) may be based on diverse criteria, including (but not restricted to), level of support, phenotypic distinctiveness, economic importance, and whether the clade has historically been named.

Another benefit of phylogenetic nomenclature is that it permits (though it does not require) the abandonment of categorical ranks, which would eliminate the most subjective aspect of traditional taxonomy. The arbitrary nature of ranking, though acknowledged by most taxonomists, is not widely appreciated by non-taxonomists. The existence of ranks encourages researchers to use taxonomies inappropriately, treating taxa at the same rank as though they were comparable in some biologically meaningful way—for example, when they count genera or families to study past and present patterns of biological diversity. A rankless system of taxonomy, which is permitted but not required by the *PhyloCode*, encourages the development of more appropriate uses of taxonomies in such studies, such as counting clades or species that possess properties relevant to the question of interest, or investigating the evolution of those properties on a phylogenetic tree.

An advantage of the *PhyloCode* over the rank-based codes is that it applies at all levels of the taxonomic hierarchy. In contrast, the zoological code does not extend its rank-based method of definition above the level of superfamily, and the botanical code extends that method of definition only to some names above the rank of family (automatically typified names) and the principle of priority is not mandatory for those names. Consequently, at higher levels in the hierarchy, the rank-based codes permit multiple names for the same taxon as well as alternative applications of the same name. Thus, as phylogenetic studies continue to reveal many deep clades, there is an increasing potential for nomenclatural chaos due to synonymy and homonymy. By imposing rules of precedence on clade names at all levels of the hierarchy, the *PhyloCode* will improve nomenclatural clarity at higher hierarchical levels.

History. The theoretical foundation of the *PhyloCode* was developed in a series of papers by de Queiroz and Gauthier (1990, 1992, 1994), which were foreshadowed by earlier suggestions that a taxon name could be defined by reference to a part of a phylogenetic tree (e.g., Ghiselin, 1984). The theory was in development for several years before the first of these theoretical papers was published, and related theoretical discussions (e.g., Rowe, 1987; de Queiroz, 1988; Gauthier et al., 1988; Estes et al., 1988) as well as explicit phylogenetic definitions (Gauthier, 1984, 1986; Gauthier and Padian, 1985; de Queiroz, 1985, 1987; Gauthier et al., 1988; Estes et al., 1988; Rowe, 1988) were published in some earlier papers. Several other papers contributed to the development of phylogenetic nomenclature prior to the Internet posting of the first version of the *PhyloCode* in 2000 (Rowe and Gauthier, 1992; Bryant, 1994, 1996, 1997; de Queiroz, 1992, 1994, 1997a, b; Sundberg and Pleijel, 1994; Christoffersen, 1995; Schander and Thollesson, 1995; Lee, 1996a, b, 1998a, b, 1999a, b; Wyss and Meng, 1996; Brochu, 1997; Cantino et al., 1997, 1999a, b; Kron, 1997; Baum et al., 1998; Cantino, 1998; Eriksson et al., 1998; Härlin, 1998, 1999; Hibbett and Donoghue, 1998; Moore, 1998; Schander, 1998a, b; Mishler, 1999; Pleijel, 1999; Sereno, 1999). Other papers during this period applied phylogenetic nomenclature to particular clades (e.g., Judd et al., 1993, 1994; Holtz, 1996; Roth, 1996; Alverson et al., 1999; Swann et al., 1999; Brochu, 1999; Bremer, 2000).

Three early symposia increased awareness of phylogenetic nomenclature. The first one, organized by Richard G. Olmstead and entitled "Translating Phylogenetic Analyses into Classifications," took place at the 1995 annual meeting of the American Institute of Biological Sciences in San Diego, California, USA. The 1996 Southwestern Botanical Systematics Symposium at the Rancho Santa Ana Botanic Garden in Claremont, California, USA, organized by J. Mark Porter and entitled "The Linnean Hierarchy: Past, Present and Future," focused in part on phylogenetic nomenclature. Philip Cantino and Torsten Eriksson organized a symposium at the XVI International Botanical Congress in St. Louis, Missouri, USA (1999), entitled "Overview and Practical Implications of Phylogenetic Nomenclature." A few critiques of phylogenetic nomenclature (Lidén and Oxelman, 1996; Dominguez and Wheeler, 1997; Lidén et al., 1997) and responses (Lee, 1996a; de Queiroz, 1997b; Schander, 1998a) were also published during this period, but the debate became much more active after the posting of the first version of the *PhyloCode* (see below).

The preparation of the *PhyloCode* began in the autumn of 1997, following a decision by Michael Donoghue, Philip Cantino, and Kevin de Queiroz to organize a workshop for this purpose. The workshop took place August 7–9, 1998, at the Harvard University Herbaria, Cambridge, Massachusetts, USA., and was attended by 27 people from five countries: William S. Alverson, Harold N. Bryant, David C. Cannatella, Philip D. Cantino, Julia Clarke, Peter R. Crane, Noel Cross, Kevin de Queiroz, Michael J. Donoghue, Torsten Eriksson, Jacques Gauthier, Kancheepuram Gandhi, Kenneth Halanych, David S. Hibbett, David M. Hillis, Kathleen A. Kron, Michael S. Y. Lee, Alessandro Minelli, Richard G. Olmstead, Fredrik Pleijel, J. Mark Porter, Heidi E. Robeck, Timothy Rowe, Christoffer Schander, Per Sundberg, Mikael Thollesson, and André R. Wyss. An initial draft of the code prepared by Cantino and de Queiroz was provided to the workshop participants in advance and was considerably revised by Cantino and de Queiroz as a result of decisions made at the meeting. The initial draft of Article 22 (Governance) was written by F. Pleijel, A. Minelli, and K. Kron and subsequently modified by M. Donoghue and P. Cantino. The initial draft of what is now Recommendation 11.10B was contributed by T.

Rowe. An earlier draft of what is now Article 10.10 was written by Gerry Moore, who also provided Example 1. Article 8 and Appendix A (both of which concern registration) were written largely by T. Eriksson. William M. Owens provided the Latin terms in Article 9.2. Whenever possible, the writers of the *PhyloCode* used as a model the draft *BioCode* (Greuter et al., 1998), which attempted to unify the rank-based approach into a single code. Thus, the organization of the *PhyloCode*, some of its terminology, and the wording of certain rules are derived from the *BioCode*. Other rules are derived from one or more of the rank-based codes, particularly the versions of the botanical and zoological codes that were in effect at that time (Greuter et al., 1994, 2000; International Commission on Zoological Nomenclature, 1985, 1999). However, many rules in the *PhyloCode* have no counterpart in any code based on taxonomic ranks because of fundamental differences in the definitional foundations of the alternative systems.

The first public draft of the *PhyloCode* was posted on the Internet in April 2000. Its existence was broadly publicized in the systematic biology community, and readers were encouraged to submit comments and suggestions. All comments received were forwarded to the advisory group via a listserver, and many of them elicited discussion. Numerous commentaries about phylogenetic nomenclature have been published since the first public posting of the *PhyloCode*, some of them critical (Benton, 2000, 2007; Nixon and Carpenter, 2000; Stuessy, 2000, 2001; Forey, 2001, 2002; Lobl, 2001; Berry, 2002; Blackwell, 2002; Jørgensen, 2002, 2004; Carpenter, 2003; Janovec et al., 2003; Keller et al., 2003; Kojima, 2003; Moore, 2003; Nixon et al., 2003; Schuh, 2003; Barkley et al., 2004; Wenzel et al., 2004; Pickett, 2005; Polaszek and Wilson, 2005; Tang and Lu, 2005; Monsch, 2006; Rieppel, 2006; Stevens, 2006; Platnick, 2009, 2012, 2013), some supportive (Bremer, 2000; Cantino, 2000, 2004; de Queiroz, 2000, 2006; Brochu and Sumrall, 2001; de Queiroz and Cantino, 2001a, b; Ereshefsky, 2001; Laurin, 2001, 2005; Lee, 2001; Bryant and Cantino, 2002; Bertrand and Pleijel, 2003; Pleijel and Rouse, 2003; Donoghue and Gauthier, 2004; Laurin, 2005a, b, 2008; Laurin et al., 2005, 2006; Lee and Skinner, 2007; de Queiroz and Donoghue, 2011, 2013), and some pointing out both advantages and disadvantages (Langer, 2001; Stevens, 2002). Other publications since 2000 have discussed properties of different kinds of phylogenetic definitions (Gauthier and de Queiroz, 2001), the application of widely used names to a particular category of clades (Anderson, 2002; Laurin, 2002; Joyce et al., 2004; Laurin and Anderson, 2004; Donoghue, 2005; Sereno, 2005), the conversion of rank-based names to phylogenetically defined names (Joyce et al., 2004), the choice of specifiers (Lee, 2005; Sereno, 2005; Wilkinson, 2006), the number of specifiers (Bertrand and Härlin, 2006), the subjective nature of Linnaean categories (Laurin, 2010), the application of phylogenetic nomenclature to species or least inclusive clades (Pleijel and Rouse, 2000, 2003; Artois, 2001; Hillis et al., 2001; Lee, 2002; Spangler, 2003; Dayrat et al., 2004; Dayrat, 2005; Dayrat and Gosliner, 2005; Fisher, 2006; Wolsan, 2007), the relevance of phylogenetic nomenclature to phyloinformatics (Donoghue, 2004; Hibbett et al., 2005), the logic and symbolic representation of phylogenetic definitions (Sereno, 2005), the philosophy of different approaches to phylogenetic nomenclature (Härlin, 2003a, b; Pleijel and Härlin, 2004), the use of phylogenetic nomenclature without a code (Sereno, 2005), guidelines for interpreting and establishing pre-*PhyloCode* phylogenetic definitions after the *PhyloCode* is implemented (Taylor, 2007), similarities between phylogenetic nomenclature and nomenclature as practiced by 18th- and early 19th century naturalists (de Queiroz, 2005, 2012), the possibility of combining elements of phylogenetic and rank-based nomenclature (Kuntner and Agnarsson, 2006), and the development of an integrated approach to naming crown and total clades (de Queiroz, 2007). There have also been many applications of phylogenetic nomenclature to particular clades (some early examples, in addition to those cited in the first paragraph of this section, are Donoghue et al., 2001; Gauthier and de Queiroz, 2001; Maryanska et al., 2002; Modesto and Anderson, 2004; Smedmark and Eriksson, 2002; Wolfe et al., 2002; Stefanovic et al., 2003; Clarke, 2004; Joyce et al., 2004; Sangster, 2005; Taylor and Naish, 2005; Cantino et al., 2007).

A second workshop on phylogenetic nomenclature was held at Yale University, July 28–30, 2002, organized by Michael Donoghue, Jacques Gauthier, Philip Cantino, and Kevin de Queiroz. There were 20 attendees from five countries, four of whom were observers. The active (voting) participants were Christopher Brochu, Harold Bryant, Philip Cantino, Kevin de Queiroz, Michael Donoghue, Torsten Eriksson, Jacques Gauthier, David Hibbett, Michel Laurin, Brent Mishler, Gerry Moore, Fredrik Pleijel, J. Mark Porter, Greg Rouse, Christoffer Schander, and Mikael Thollesson. Sixteen proposed changes in the rules and recommendations were discussed, 11 of which were approved. (Many other minor wording changes had already been circulated by e-mail and approved in advance of the workshop.)

In addition to specific rule changes, the 2002 workshop focused on several larger issues, the most fundamental of which concerned the governance of species names. The first public draft of the *PhyloCode* covered only clade names. Among the advisory group members, there was a diversity of opinions on how species names should be handled, ranging from those who thought that species names should never be governed by the *PhyloCode* to those who argued that their inclusion is so essential that the *PhyloCode* should not be implemented until rules governing species names have been added. The majority held the intermediate view that species names should eventually be included in the *PhyloCode* but that the controversy surrounding species and species names, both within the advisory group and in the systematics community as a whole, should not be allowed to delay implementation of the rules for clade names. Thus, it was decided, first, that rules for clade names and rules for species names would be published in separate documents and, second, that the timing of implementation of the two documents would be independent; thus, the rules for clade names would likely be implemented before those for species names. (This decision was reconsidered in 2006, and a different approach to species names was adopted by the CPN in 2007; see below.)

A second major decision at the 2002 Yale workshop concerned the proposal of a companion volume, to be published simultaneously with the *PhyloCode*, that would define various clade names following the rules of the *PhyloCode* and serve as its starting point with regard to priority. (This companion volume was later named *Phylonyms: A Companion to the PhyloCode*.) As originally conceived, the companion volume would have included phylogenetic definitions of the most widely known names in most major groups of organisms. It was soon realized that several volumes would be needed, that producing these volumes would be an immense job, and that linking the starting date of the *PhyloCode* to their publication would greatly delay its implementation. For this reason, the participants in the second workshop decided to reduce the scope of the companion volume. Instead of attempting a comprehensive treatment of widely known clade names for all major groups of organisms, the companion volume would include only examples involving taxa for which there were systematists who could be recruited to contribute entries. A plan for a conference was conceived in which participants would apply phylogenetic nomenclature to clades that they study. The definitions from the papers presented at the conference would form the nucleus of the companion volume. Michel Laurin offered to organize the meeting, and Kevin de Queiroz and Jacques Gauthier were chosen to edit the companion volume (Philip Cantino was enlisted in 2004 as a third editor).

The First International Phylogenetic Nomenclature Meeting took place July 6–9, 2004, at the Muséum National d'Histoire Naturelle in Paris, organized by a 10-member committee chaired by Michel Laurin. The meeting was described in detail by Laurin and Cantino (2004), and the program and abstracts are available at www.phylocode.org and www.phylonames.org. Unlike the preceding workshops, this conference included research presentations and was open to anyone interested in attending. It was attended by 70 people from 11 countries, and 36 papers were presented. The Paris conference also served as the inaugural meeting of the International Society for Phylogenetic Nomenclature (ISPN), including the election of a governing council and officers and approval of the bylaws (available at the subsequently established ISPN website: www.phylonames. org). The ISPN includes an elected Committee on Phylogenetic Nomenclature, whose responsibilities include ratifying the first edition of the *PhyloCode* and approving any subsequent modifications (for full responsibilities, see Article 22).

Presentations were given at the Paris meeting on the theory and practice of phylogenetic nomenclature and its applications to a wide variety of groups. Besides the inauguration of the ISPN, there were several other important outcomes of the meeting: (1) A proposal by K. de Queiroz and J. Gauthier to adopt "an integrated system of phylogenetically defined names," including the application of widely known names to crown clades and forming the names of the corresponding total clades by adding the prefix "*Pan-*" to the name of the crown (Lauterbach, 1989; Meier and Richter, 1992; Gauthier and de Queiroz, 2001; Joyce et al., 2004), was introduced and vigorously discussed. Some participants were reluctant to make these conventions mandatory because doing so would result in replacing some names that had already been explicitly defined as the names of total clades (e.g., replacing *Synapsida* by *Pan-Mammalia*). A compromise that made exceptions for such names was acceptable to the majority of the participants, and it served as the basis for the set of rules and recommendations that was eventually adopted by the CPN (Recommendation 10.1B and Articles 10.3–10.8 in version 3 of the *PhyloCode*, and after some subsequent modifications, Recommendation 10.1B and Articles 10.3–10.7 in the current version). (2) Benoît Dayrat proposed that phylogenetically defined species names consist of a single word (the second part of the binomen in the case of already existing names) plus the author of the name, year of publication, and (if necessary to ensure uniqueness) the page number where published (Dayrat et al., 2004). In practice, the name of a small clade (generally corresponding to a genus under the rank-based system) would likely be cited before the species name, but it would not be part of the species name. In conversation and in teaching, the name would likely be abbreviated to the single word (i.e., omitting the author and year) when doing so is unambiguous. Dayrat's proposal was well received by conference participants. (3) Julia Clarke proposed a flexible way of defining species names that is applicable to the wide variety of entities that are called species. The definitions would take the form "the species that includes specimen X" (de Queiroz, 1992), and the author would be required to explain what he/she means by "species." This approach is similar to the way species names are implicitly defined in rank-based nomenclature but differs in that the species category is not a rank, and the author is required to explain the kind of entity to which the name refers. (4) In a straw vote of meeting participants, it was decided that Clarke, Dayrat, Cantino, and de Queiroz would draft a code for species names based on the above-described proposals of Clarke and Dayrat. Consistent with the decision made at the 2002 Yale workshop, this code would be separate from, but compatible with, the code for clade names.

In the fall of 2004, Cantino and de Queiroz drafted a code for species names based on the proposals approved at the Paris meeting. After review of the draft by Dayrat and Clarke and e-mail discussion of unresolved issues, the four potential authors of the code met at the Smithsonian Institution on May 20–21, 2006. In the process of drafting the code, the seriousness of the drawbacks of extending the *PhyloCode* to species names using an epithet-based format had become more apparent. Most critically, species names would be different under rank-based and phylogenetic nomenclature (e.g., "*Homo sapiens*" vs. "*sapiens* Linnaeus 1758"), which would create confusion. Second, differences in the way types are handled under the zoological and botanical codes would complicate the development of a universal code governing species names. Third, establishing and registering reformatted names for every species known to science would be an immense job—and one of questionable value given that there would be no fundamental difference in the way that the names would be defined. What emerged from the May 2006 meeting was an entirely different (and much simpler) way to reconcile the incompatibilities between traditional binominal species names and phylogenetic nomenclature—including the mandatory genus category and the fact that many genera are not monophyletic. This approach was subsequently adopted by the CPN (May, 2007), described in detail by Dayrat et al. (2008), and incorporated into the code (as Article 21).

The Second Meeting of the ISPN took place June 28–July 2, 2006, at the Peabody Museum of Yale University, organized by an eight-member committee co-chaired by Nico Cellinese and Walter Joyce. Most of the papers were presented in three symposia: phylogenetic nomenclature of species (organized by David Baum and Benoît Dayrat), implementing phylogenetic nomenclature (organized by Philip Cantino), and phyloinformatics (organized by Michael Donoghue and Nico Cellinese). The meeting was described in detail by Laurin and Cantino (2006, 2007), and the program and abstracts are available at www.phylocode.org and www.phylonames.org. At this second ISPN meeting, more time was devoted to open-ended discussions of issues raised in the presentations. The new approach to species names that was developed at the May 2006 meeting in Washington (see previous paragraph) was presented in talks by Clarke and Dayrat and was well received in the subsequent discussion. They and their coauthors (Cantino and de Queiroz) were encouraged to continue work on a set of rules and recommendations that would implement this approach.

Another issue that generated a lot of discussion at the second ISPN meeting was the integrated system of crown- and total-clade names that was introduced at the 2004 Paris meeting and incorporated into *PhyloCode* version 3. Although the rules and recommendations promoting an integrated system in version 3 represented a compromise, there was still a lot of dissatisfaction on the part of some discussants. An alternative means of referring to total clades using "pan" as a function name was proposed by T. Michael Keesey. In the course of the discussion, it was suggested that the prefix "pan-" (lower case) be used to designate informal names for total clades that may or may not have a formal name. Because informal names do not compete with formal names for precedence, they can coexist without violating Principle 3 (that each taxon may have only one accepted name). Using this approach, a widely used name could be retained for a total clade and coexist with an informal name with the prefix "pan-." For example, the total clade of mammals might have the formal name *Synapsida* and the informal name pan-Mammalia. This suggestion led to changes in Article 10 that were approved by the CPN in January 2007 and included in this version of the code.

The Third Meeting of the ISPN took place July 20–22, 2008, at Dalhousie University in Halifax, Nova Scotia, organized by a four-member committee chaired by Harold Bryant. It was coordinated with a joint meeting of the International Society of Protistologists and the International Society for Evolutionary Protistology. In addition to contributed papers (see Laurin and Bryant, 2009, for details), including a plenary talk by Sina Adl focusing on issues in protist nomenclature, the meeting focused on how to expedite completion of two critical projects that must reach fruition before implementation of the code: preparation of the companion volume (*Phylonyms*), to be published simultaneously with the code, and development of the registration database (RegNum).

Because registration is required for establishment of names under the *PhyloCode*, the registration database (which has come to be known as RegNum) had to be developed before the code could be implemented. Torsten Eriksson and Mikael Thollesson initially designed the database structure and reported on it at the 2002 Yale workshop. Further development of the database and web/user interface was subsequently carried out at Uppsala University by Jonas Ekstedt and M. Thollesson. An alpha test site for this version was announced at the 2004 Paris meeting. A prototype of RegNum was demonstrated at a meeting of the ISPN Registration Committee (Mikael Thollesson, Torsten Eriksson, and Nico Cellinese) and other interested persons at Yale University on November 2–3, 2005, and it was subsequently demonstrated at both the 2006 and 2008 ISPN meetings. In 2009, Nico Cellinese (University of Florida), who was chair of the Registration Committee at the time of this writing, started the development of a more comprehensive and flexible prototype. The new version of RegNum was conceived in line with other developments concerning biological name repositories and resolution services (e.g., Global Name Architecture). RegNum's prototype was demonstrated at the 2009 Biodiversity Information Standards (TDWG) meeting in Montpellier, France, during a workshop that focused on phylogenetic nomenclature and future informatics development. The RegNum database currently satisfies the requirements of the *PhyloCode*, and plans are in place to integrate it with several tools and data repositories (e.g., TreeBASE) that will enhance its relevance for phylogenetic research.

In October 2011, the CPN received a proposal by Nico Cellinese, David Baum, and Brent Mishler concerning the treatment of species in this code, the justification for which was published by Cellinese et al. (2012). Their fundamental premise was that the code is too strongly tied to a particular view on the nature of species, which is not accepted by everyone who would like to use phylogenetic nomenclature, and their proposed solution was to eliminate all mention of species in the code. The proposal was publicized on the ISPN website and stimulated more than a year of intermittent discussion on the CPN listserv. In the end, the CPN accepted the underlying premise of the proposal but rejected the proposed solution, which would have entailed radical changes in the code, including: eliminating the use of species as specifiers (thus only specimens would be specifiers); permitting the establishment of preexisting specific epithets as clade names; redefining the term "homonym" such that established clade names may be identical provided that the authors and publication years differ; and elimination of Article 21, which provides recommendations on how to use species names governed by the rank-based codes in conjunction with clade names governed by this code. Instead of eliminating all reference to species, the definition of "species" was broadened to accommodate the view that the species category is simply a rank in the Linnaean hierarchy, while continuing to accommodate the view that it is a kind of biological entity. In addition, the CPN discussion of the Cellinese et al. proposal led to other changes not proposed by those authors, especially changes in Article 11 related to using species versus type specimens as specifiers.

An important change approved by the CPN in August 2013 was the expansion of the treatment of phylogenetic definitions, which in many ways are the heart of this code. What were formerly a long and complicated note (9.3.1) and three related recommendations (9.3B, 9D, and 9E) were expanded into six articles (9.4–9.7 and 9.9–9.10). Following a proposal by de Queiroz (2013) that was motivated by a distinction emphasized by Martin et al. (2010), node-based and branch-based definitions were replaced with minimum-clade and maximum-clade definitions, respectively, throughout the code. The primary reason for this change was to employ definitions that are applicable in the context of both common interpretations of phylogenetic trees—one in which branches represent lineages and nodes represent ancestors at lineage-splitting events; the other in which branches represent relationships and nodes represent taxa (the term "branch-based" is inappropriate in the latter context, in which all phylogenetic definitions are effectively node-based).

Another important change in Article 9 is the explicit recognition of definition categories for the names of crown clades (Article 9.9) and total clades (Article 9.10), the variants of which (e.g., the maximum-crown-clade definition, which is roughly equivalent to what were previously called branch-modified node-based definition) are presented more systematically and thoroughly than in previous versions of the code. Additionally, a new note (9.5.1) was added describing a kind of minimum-clade definition that had no analog in previous versions of this code, the directly-specified-ancestor definition, wherein the ancestor in which the clade originated is identified by name rather than being specified indirectly through its descendants, and Note 9.9.2 was added to describe its use for defining the names of crown clades.

Readers interested in more information about the sequence of changes from one version of this code to the next are referred to www.phylocode.org, where all previous versions are available. The changes implemented in versions 3 and 4 are summarized in the preface of each, and changes implemented in version 5 are summarized in a separate document.

There is only one major modification in the current version of the code (version 6) relative to version 5. The rules on publication (Articles 4 and 5) have been revised to permit electronic publication, based on modifications proposed by Nico Cellinese and Richard Olmstead. Most other changes from version 5 are simply clarifications, but two new rules (Articles 10.7 and 14.5) have been added to ensure that accepted panclade names are always based on the names of the corresponding crown clades.

Acknowledgments. We thank current and past members of the Committee on Phylogenetic Nomenclature for reviewing and approving earlier drafts of the Code as well as proposed additions, deletions, and other modifications: Sina Adl, Frank (Andy) Anderson, Brian Andres, Tom Artois, Christopher Brochu, Harold Bryant, David Cannatella, Nico Cellinese, Julia Clarke, Benoît Dayrat, Jim Doyle, Micah Dunthorn, Jacques Gauthier, Sean Graham, John Hall, David Hibbett, David Hillis, Walter Joyce, Michael Keesey, Max Cardoso Langer, Michel Laurin, David Marjanović, Richard Olmstead, Kevin Padian, Fredrik Pleijel, Greg Rouse, George Sangster, David Tank, and Mieczyslaw Wolsan. We wish to give special thanks to David Marjanović for his thorough reviews and numerous constructive suggestions, Michael Keesey for design and maintenance of the PhyloCode website, and Michael Donoghue for the initial push to undertake this project. Chuck Crumly has tirelessly and patiently supported publication of this Code for many years. We also wish to thank Michele Dimont at CRC Press and Rachel Cook at Deanta Global for overseeing copy-editing and production of the book. Other people who have made important contributions to this project are mentioned in the History section above.

LITERATURE CITED

Alverson, W. S., B. A. Whitlock, R. Nyffeler, C. Bayer, and D. A. Baum. 1999. Phylogeny of the core *Malvales*: evidence from *ndhF* sequence data. *Am. J. Bot.* 86:1474–1486.

Anderson, J. S. 2002. Use of well-known names in phylogenetic nomenclature: a reply to Laurin. *Syst. Biol.* 51:822–827.

Artois, T. 2001. Phylogenetic nomenclature: the end of binomial nomenclature? *Belg. J. Zool.* 131:87–89.

Barkley, T. M., P. DePriest, V. Funk, R. W. Kiger, W. J. Kress, and G. Moore. 2004. Linnaean nomenclature in the 21st century: a report from a workshop on integrating traditional nomenclature and phylogenetic classification. *Taxon* 53:153–158.

Baum, D. A., W. S. Alverson, and R. Nyffeler. 1998. A durian by any other name: taxonomy and nomenclature of the core *Malvales*. *Harv. Pap. Bot.* 3:315–330.

Benton, M. J. 2000. Stems, nodes, crown clades, and rank-free lists: is Linnaeus dead? *Biol. Rev.* 75:633–648.

Benton, M. J. 2007. The Phylocode: beating a dead horse? *Acta Paleontol. Pol.* 52:651–655.

Berry, P. E. 2002. Biological inventories and the PhyloCode. *Taxon* 51:27–29.

Bertrand, Y., and M. Härlin. 2006. Stability and universality in the application of taxon names in phylogenetic nomenclature. *Syst. Biol.* 55:848–858.

Bertrand, Y., and F. Pleijel. 2003. Nomenclature phylogénétique: une reponse. *Bull. Soc. Fr. Syst.* 29:25–28.

Blackwell, W. H. 2002. One-hundred-year code déjà vu? *Taxon* 51:151–154.

Bremer, K. 2000. Phylogenetic nomenclature and the new ordinal system of the angiosperms. Pages 125–133 *in "Plant Systematics for the 21st Century"* (B. Nordenstam, G. El-Ghazaly and M. Kassas, eds.). Portland Press, London.

Brochu, C. A. 1997. Synonymy, redundancy, and the name of the crocodile stem-group. *J. Vertebr. Paleontol.* 17:448–449.

Brochu, C. A. 1999. Phylogenetics, taxonomy, and historical biogeography of *Alligatoroidea. J. Vertebr. Paleontol.* 19 (suppl. to no. 2):9–100.

Brochu, C. A., and C. D. Sumrall. 2001. Phylogenetic nomenclature and paleontology. *J. Paleontol.* 75:754–757.

Bryant, H. N. 1994. Comments on the phylogenetic definition of taxon names and conventions regarding the naming of crown clades. *Syst. Biol.* 43:124–130.

Bryant, H. N. 1996. Explicitness, stability, and universality in the phylogenetic definition and usage of taxon names: a case study of the phylogenetic taxonomy of the *Carnivora (Mammalia). Syst. Biol.* 45:174–189.

Bryant, H. N. 1997. Cladistic information in phylogenetic definitions and designated phylogenetic contexts for the use of taxon names. *Biol. J. Linn. Soc.* 62:495–503.

Bryant, H. N., and P. D. Cantino. 2002. A review of criticisms of phylogenetic nomenclature: is taxonomic freedom the fundamental issue? *Biol. Rev.* 77:39–55.

Cantino, P. D. 1998. Binomials, hyphenated uninomials, and phylogenetic nomenclature. *Taxon* 47:425–429.

Cantino, P. D. 2000. Phylogenetic nomenclature: addressing some concerns. *Taxon* 49:85–93.

Cantino, P. D. 2004. Classifying species versus naming clades. *Taxon* 53:795–798.

Cantino, P. D., H. N. Bryant, K. de Queiroz, M. J. Donoghue, T. Eriksson, D. M. Hillis, and M. S. Y. Lee. 1999b. Species names in phylogenetic nomenclature. *Syst. Biol.* 48:790–807.

Cantino, P. D., J. A. Doyle, S. W. Graham, W. S. Judd, R. G. Olmstead, D. E. Soltis, P. S. Soltis, and M. J. Donoghue. 2007. Towards a phylogenetic nomenclature of *Tracheophyta. Taxon* 56:822–846.

Cantino, P. D., R. G. Olmstead, and S. J. Wagstaff. 1997. A comparison of phylogenetic nomenclature with the current system: a botanical case study. *Syst. Biol.* 46:313–331.

Cantino, P. D., S. J. Wagstaff, and R. G. Olmstead. 1999a. *Caryopteris* (*Lamiaceae*) and the conflict between phylogenetic and pragmatic considerations in botanical nomenclature. *Syst. Bot.* 23:369–386.

Carpenter, J. M. 2003. Critique of pure folly. *Bot. Rev.* 69:79–92.

Cellinese, N., D. A. Baum, and B. D. Mishler. 2012. Species and phylogenetic nomenclature. *Syst. Biol.* 61:885–891.

Christoffersen, M. L. 1995. Cladistic taxonomy, phylogenetic systematics, and evolutionary ranking. *Syst. Biol.* 44:440–454.

Clarke, J. A. 2004. Morphology, phylogenetic taxonomy, and systematics of *Ichthyornis* and *Apatornis* (*Avialae: Ornithurae*). *Bull. Am. Mus. Nat. Hist.* 286:1–179.

Dayrat, B. 2005. Advantages of naming species under the PhyloCode: an example of how a new species of *Discodorididae* (*Mollusca, Gastropoda, Euthyneura, Nudibranchia, Doridina*) may be named. *Mar. Biol. Res.* 1:216–232.

Dayrat, B., P. D. Cantino, J. A. Clarke, and K. de Queiroz. 2008. Species names in the *PhyloCode*: the approach adopted by the International Society for Phylogenetic Nomenclature. *Syst. Biol.* 57:507–514.

Dayrat, B., and T. M. Gosliner. 2005. Species names and metaphyly: a case study in *Discodorididae* (*Mollusca, Gastropoda, Euthyneura, Nudibranchia, Doridina*). *Zool. Scr.* 34:199–224.

Dayrat, B., C. Schander, and K. D. Angielczyk. 2004. Suggestions for a new species nomenclature. *Taxon* 53:485–591.

de Queiroz, K. 1985. *Phylogenetic systematics of iguanine lizards: a comparative osteological study.* Master's thesis, San Diego State University, San Diego, CA.

de Queiroz, K. 1987. Phylogenetic systematics of iguanine lizards. A comparative osteological study. *Univ. Calif. Publ. Zool.* 118:1–203.

de Queiroz, K. 1988. Systematics and the Darwinian revolution. *Philos. Sci.* 55:238–259.

de Queiroz, K. 1992. Phylogenetic definitions and taxonomic philosophy. *Biol. Philos.* 7:295–313.

de Queiroz, K. 1994. Replacement of an essentialistic perspective on taxonomic definitions as exemplified by the definition of "*Mammalia*." *Syst. Biol.* 43:497–510.

de Queiroz, K. 1997a. The Linnaean hierarchy and the evolutionization of taxonomy, with emphasis on the problem of nomenclature. *Aliso* 15:125–144.

de Queiroz, K. 1997b. Misunderstandings about the phylogenetic approach to biological nomenclature: a reply to Lidén and Oxelman. *Zool. Scr.* 26:67–70.

de Queiroz, K. 2000. The definitions of taxon names: a reply to Stuessy. *Taxon* 49:533–536.

de Queiroz, K. 2005. Linnaean, rank-based, and phylogenetic nomenclature: restoring primacy to the link between names and taxa. *Symbolae Botanicae Upsaliensis* 33:127–140.

de Queiroz, K. 2006. The PhyloCode and the distinction between taxonomy and nomenclature. *Syst. Biol.* 55:160–162.

de Queiroz, K. 2007. Toward an integrated system of clade names. *Syst. Biol.* 56:956–974.

de Queiroz, K. 2012. Biological nomenclature from Linnaeus to the PhyloCode. *Bibliotheca Herpetologica* 9:135–145.

de Queiroz, K. 2013. Nodes, branches, and phylogenetic definitions. *Syst. Biol.* 62:625–632.

de Queiroz, K., and P. D. Cantino. 2001a. Phylogenetic nomenclature and the PhyloCode. *Bull. Zool. Nomencl.* 58:254–271.

de Queiroz, K., and P. D. Cantino. 2001b. Taxon names, not taxa, are defined. *Taxon* 50:821–826.

de Queiroz, K., and M. J. Donoghue. 2011. Phylogenetic nomenclature, three-taxon statements, and unnecessary name changes. *Syst. Biol.* 60:887–892.

de Queiroz, K., and M. J. Donoghue. 2013. Phylogenetic nomenclature, hierarchical information, and testability. *Syst. Biol.* 62:167–174.

de Queiroz, K., and J. Gauthier. 1990. Phylogeny as a central principle in taxonomy: phylogenetic definitions of taxon names. *Syst. Zool.* 39:307–322.

de Queiroz, K., and J. Gauthier. 1992. Phylogenetic taxonomy. *Annu. Rev. Ecol. Syst.* 23:449–480.

de Queiroz, K., and J. Gauthier. 1994. Toward a phylogenetic system of biological nomenclature. *Trends Ecol. Evol.* 9:27–31.

Dominguez, E., and Q. D. Wheeler. 1997. Taxonomic stability is ignorance. *Cladistics* 13:367–372.

Donoghue, M. J. 2004. Immeasurable progress on the tree of life. Pages 548–552 *in "Assembling the tree of life"* (J. Cracraft and M. J. Donoghue, eds.). Oxford University Press, Oxford, UK.

Donoghue, M. J., T. Eriksson, P. A. Reeves, and R. G. Olmstead. 2001. Phylogeny and phylogenetic taxonomy of *Dipsacales*, with special reference to *Sinadoxa* and *Tetradoxa* (*Adoxaceae*). *Harvard Pap. Bot.* 6:459–479.

Donoghue, M. J., and J. A. Gauthier. 2004. Implementing the PhyloCode. *Trends Ecol. Evol.* 19:281–282.

Donoghue, P. C. J. 2005. Saving the stem group—a contradiction in terms? *Paleobiology* 31:553–558.

Ereshefsky, M. 2001. *The poverty of the Linnaean hierarchy: a philosophical study of biological taxonomy.* Cambridge University Press, Cambridge, UK.

Eriksson, T., M. J. Donoghue, and M. S. Hibbs. 1998. Phylogenetic analysis of *Potentilla* using DNA sequences of nuclear ribosomal internal transcribed spacers (ITS), and implications for the classification of *Rosoideae* (*Rosaceae*). *Plant Syst. Evol.* 211:155–179.

Estes, R., K. de Queiroz, and J. Gauthier. 1988. Phylogenetic relationships within *Squamata*. Pages 119–281 *in "Phylogenetic relationships of the lizard families: essays commemorating Charles L. Camp"* (R. Estes and G. K. Pregill, eds.). Stanford University Press, Stanford, CA.

Fisher, K. 2006. Rank-free monography: a practical example from the moss clade *Leucophanella* (*Calymperaceae*). *Syst. Bot.* 31:13–30.

Forey, P. L. 2001. The PhyloCode: description and commentary. *Bull. Zool. Nomencl.* 58:81–96.

Forey, P. L. 2002. PhyloCode—pain, no gain. *Taxon* 51:43–54.

Gauthier, J. 1984. *A cladistic analysis of the higher systematic categories of the Diapsida.* Ph.D. dissertation, University of California at Berkeley, Berkeley, CA.

Gauthier, J. 1986. Saurischian monophyly and the origin of birds. Pages 1–55 *in "The origin of birds and the evolution of flight"* (K. Padian, ed.). California Academy of Sciences, San Francisco, CA.

Gauthier, J., and K. de Queiroz. 2001. Feathered dinosaurs, flying dinosaurs, crown dinosaurs, and the name "*Aves*". Pages 7–41 *in "New perspectives on the origin and early evolution of birds: proceedings of the*

International Symposium in Honor of John H. Ostrom" (J. Gauthier and L. F. Gall, eds.). Peabody Museum of Natural History, Yale University, New Haven, CT.

Gauthier, J., R. Estes, and K. de Queiroz. 1988. A phylogenetic analysis of *Lepidosauromorpha*. Pages 15–98 *in* "*Phylogenetic relationships of the lizard families: essays commemorating Charles L. Camp*" (R. Estes and G. K. Pregill, eds.). Stanford University Press, Stanford, CA.

Gauthier, J., and K. Padian. 1985. Phylogenetic, functional, and aerodynamic analyses of the origin of birds and their flight. Pages 185–197 *in* "*The beginnings of birds*" (M. K. Hecht, J. H. Ostrom, G. Viohl. and P. Wellnhofer, eds.). Freunde des Jura-Museums, Eichstatt, Germany.

Ghiselin, M. T. 1984. "Definition," "character," and other equivocal terms. *Syst. Zool.* 33:104–110.

Greuter, W., F. R. Barrie, H. M. Burdet, W. G. Chaloner, V. Demoulin, D. L. Hawksworth, P. M. Jørgensen, J. McNeill, D. H. Nicolson, P. C. Silva, and P. Trehane. 1994. *International code of botanical nomenclature (Tokyo code)*. Koeltz Scientific Books, Königstein, Germany.

Greuter, W., F. R. Barrie, H. M. Burdet, V. Demoulin, T. S. Filgueiras, D. L. Hawksworth, J. McNeill, D. H. Nicolson, P. C. Silva, J. E. Skog, P. Trehane, and N. J. Turland. 2000. *International code of botanical nomenclature (Saint Louis code)*. Koeltz Scientific Books, Königstein, Germany.

Greuter, W., D. L. Hawksworth, J. McNeill, M. A. Mayo, A. Minelli, P. H. A. Sneath, B. J. Tindall, P. Trehane, and P. Tubbs. 1998. Draft BioCode (1997): the prospective international rules for the scientific names of organisms. *Taxon* 47:127–150.

Härlin, M. 1998. Taxonomic names and phylogenetic trees. *Zool. Scr.* 27:381–390.

Härlin, M. 1999. The logical priority of the tree over characters and some of its consequences for taxonomy. *Biol. J. Linn. Soc.* 68:497–503.

Härlin, M. 2003a. Taxon names as paradigms: the structure of nomenclatural revolutions. *Cladistics* 19:138–143.

Härlin, M. 2003b. On the relationship between content, ancestor, and ancestry in phylogenetic nomenclature. *Cladistics* 19:144–147.

Hibbett, D. S., and M. J. Donoghue. 1998. Integrating phylogenetic analysis and classification in fungi. *Mycologia* 90:347–356.

Hibbett, D. S., R. H. Nilsson, M. Snyder, M. Fonseca, J. Costanzo, and M. Shonfeld. 2005. Automated phylogenetic taxonomy: an example in the *Homobasidiomycetes* (mushroom-forming fungi). *Syst. Biol.* 54:660–668.

Hillis, D. M., D. A. Chamberlain, T. P. Wilcox, and P. T. Chippindale. 2001. A new species of subterranean blind salamander (*Plethodontidae*: *Hemidactyliini*: *Eurycea*: *Typhlomolge*) from Austin, Texas, and a systematic revision of central Texas paedomorphic salamanders. *Herpetologica* 57:266–280.

Holtz, T. R. 1996. Phylogenetic taxonomy of the *Coelurosauria* (*Dinosauria*: *Theropoda*). *J. Paleontol.* 70:536–538.

International Commission on Zoological Nomenclature. 1985. *International code of zoological nomenclature*, 3rd ed. International Trust for Zoological Nomenclature.

International Commission on Zoological Nomenclature. 1999. *International code of zoological nomenclature*, 4th ed. International Trust for Zoological Nomenclature.

Janovec, J. P., L. G. Clark, and S. A. Mori. 2003. Is the neotropical flora ready for the PhyloCode? *Bot. Rev.* 69:22–43.

Jørgensen, P. M. 2002. Two nomenclatural systems? *Taxon* 51:737.

Jørgensen, P. M. 2004. Rankless names in the *Code*? *Taxon* 53:162.

Joyce, W. G., J. F. Parham, and J. A. Gauthier. 2004. Developing a protocol for the conversion of rank-based taxon names to phylogenetically defined clade names, as exemplified by turtles. *J. Paleontol.* 78:989–1013.

Judd, W. S., R. W. Sanders, and M. J. Donoghue. 1994. Angiosperm family pairs: preliminary phylogenetic analyses. *Harvard Pap. Bot.* 5:1–51.

Judd, W. S., W. L. Stern, and V. I. Cheadle. 1993. Phylogenetic position of *Apostasia* and *Neuwiedia* (*Orchidaceae*). *Bot. J. Linn. Soc.* 113:87–94.

Keller, R. A., R. N. Boyd, and Q. D. Wheeler. 2003. The illogical basis of phylogenetic nomenclature. *Bot. Rev.* 69:93–110.

Kojima, J. 2003. Apomorphy-based definition also pinpoints a node, and PhyloCode names prevent effective communication. *Bot. Rev.* 69:44–58.

Kron, K. A. 1997. Exploring alternative systems of classification. *Aliso* 15:105–112.

Kuntner, M., and I. Agnarsson. 2006. Are the Linnean and phylogenetic nomenclatural systems combinable? Recommendations for biological nomenclature. *Syst. Biol.* 55:774–784.

Langer, M. C. 2001. Linnaeus and the PhyloCode: where are the differences? *Taxon* 50:1091–1096.

Laurin, M. 2001. L'utilisation de la taxonomie phylogénétique en paléontologie: avatages et inconvénients. *Biosystema (Systématique et Paléontologie)* 19:197–211.

Laurin, M. 2002. Tetrapod phylogeny, amphibian origins, and the definition of the name *Tetrapoda*. *Syst. Biol.* 51:364–369.

Laurin, M. 2005a. Dites oui au PhyloCode. *Bull. Soc. Fr. Syst.* 34:25–31.

Laurin, M. 2005b. The advantages of phylogenetic nomenclature over Linnean nomenclature. Pages 67–97 *in* "*Animal names*" (A. Minelli, G. Ortalli, and G. Sanga, eds.). Istituto Veneto di Scienze, Lettere ed Arti, Venice.

Laurin, M. 2008. The splendid isolation of biological nomenclature. *Zool. Scr.* 37:223–233.

Laurin, M. 2010. The subjective nature of Linnaean categories and its impact in evolutionary biology and biodiversity studies. *Contrib. Zool.* 79:131–146.

Laurin, M., and J. S. Anderson. 2004. Meaning of the name *Tetrapoda* in the scientific literature: an exchange. *Syst. Biol.* 53:68–80.

Laurin, M., and H. N. Bryant. 2009. Third meeting of the International Society for Phylogenetic Nomenclature: a report. *Zool. Scr.* 38:333–337.

Laurin, M., and P. D. Cantino. 2004. First International Phylogenetic Nomenclature Meeting: a report. *Zool. Scr.* 33:475–479.

Laurin, M., and P. D. Cantino. 2006. Second Congrès International de la Société de Nomenclature Phylogénétique: 28 juin–2 juillet, 2006, Université de Yale, USA. *J. Assoc. Paléontol. Française* 50:18–21.

Laurin, M., and P. D. Cantino. 2007. Second meeting of the International Society for Phylogenetic Nomenclature: a report. *Zool. Scr.* 36:109–117.

Laurin, M., K. de Queiroz, P. Cantino, N. Cellinese, and R. Olmstead. 2005. The PhyloCode, types, ranks, and monophyly: a response to Pickett. *Cladistics* 21:605–607.

Laurin, M., K. de Queiroz, and P. D. Cantino. 2006. Sense and stability of taxon names. *Zool. Scr.* 35:113–114.

Lauterbach, K.-E. 1989. Das Pan-Monophylum—Ein Hilfsmittel für die Praxis der phylogenetischen Systematik. *Zool. Anz.* 223:139–156.

Lee, M. S. Y. 1996a. The phylogenetic approach to biological taxonomy: practical aspects. *Zool. Scr.* 25:187–190.

Lee, M. S. Y. 1996b. Stability in meaning and content of taxon names: an evaluation of crown-clade definitions. *Proc. R. Soc. Lond. B Biol. Sci.* 263:1103–1109.

Lee, M. S. Y. 1998a. Phylogenetic uncertainty, molecular sequences, and the definition of taxon names. *Syst. Biol.* 47:719–726.

Lee, M. S. Y. 1998b. Ancestors and taxonomy. *Trends Ecol. Evol.* 13:26.

Lee, M. S. Y. 1999a. Reference taxa and phylogenetic nomenclature. *Taxon* 48:31–34.

Lee, M. S. Y. 1999b. Stability of higher taxa in phylogenetic nomenclature—some comments on Moore (1998). *Zool. Scr.* 28:361–366.

Lee, M. S. Y. 2001. On recent arguments for phylogenetic nomenclature. *Taxon* 50:175–180.

Lee, M. S. Y. 2002. Species and phylogenetic nomenclature. *Taxon* 51:507–510.

Lee, M. S. Y. 2005. Choosing reference taxa in phylogenetic nomenclature. *Zool. Scr.* 34:329–331.

Lee, M. S. Y., and A. W. Skinner. 2007. Stability, ranks, and the PhyloCode. *Acta Palaeontol. Pol.* 52:643–650.

Lidén, M., and B. Oxelman. 1996. Do we need phylogenetic taxonomy? *Zool. Scr.* 25:183–185.

Lidén, M., B. Oxelman, A. Backlund, L. Andersson, B. Bremer, R. Eriksson, R. Moberg, I. Nordal, K. Persson, M. Thulin, and B. Zimmer. 1997. Charlie is our darling. *Taxon* 46:735–738.

Lobl, I. 2001. Les nomenclatures "linéenne" et "phylogénetique", et d'autres problèmes artificiels. *Bull. Soc. Fr. Syst.* 26:16–21.

Martin, J., D. Blackburn, and E. O. Wiley. 2010. Are node-based and stem-based clades equivalent? Insights from graph theory. *PLOS Curr. Tree of Life* 2:1–12.

Maryanska, T., H. Osmólska, and M. Wolsan. 2002. Avialan status for Oviraptorosauria. *Acta Palaeontol. Pol.* 47:97–116.

Meier, R., and S. Richter. 1992. Suggestions for a more precise usage of proper names of taxa: ambiguities related to the stem lineage concept. *Z. Zool. Syst. Evol.* 30:81–88.

Mishler, B. D. 1999. Getting rid of species? Pages 307–315 *in* "*Species: new interdisciplinary essays*" (R. Wilson, ed.). M.I.T. Press, Cambridge, MA.

Modesto, S. P., and J. S. Anderson. 2004. The phylogenetic definition of *Reptilia*. *Syst. Biol.* 53:815–821.

Monsch, K. A. 2006. The PhyloCode, or alternative nomenclature: why it is not beneficial to palaeontology, either. *Acta Palaeontol. Pol.* 51:521–524.

Moore, G. 1998. A comparison of traditional and phylogenetic nomenclature. *Taxon* 47:561–579.

Moore, G. 2003. Should taxon names be explicitly defined? *Bot. Rev.* 69:2–21.

Nixon, K. C., and J. M. Carpenter. 2000. On the other "phylogenetic systematics". *Cladistics* 16:298–318.

Nixon, K. C., J. M. Carpenter, and D. W. Stevenson. 2003. The PhyloCode is fatally flawed, and the "Linnaean" system can easily be fixed. *Bot. Rev.* 69:111–120.

Pickett, K. M. 2005. The new and improved PhyloCode, now with types, ranks, and even polyphyly: a conference report from the First International Phylogenetic Nomenclature Meeting. *Cladistics* 21:79–82.

Platnick, N. I. 2009. [Letter to Linnaeus]. Pages 199–203 *in* "*Letters to Linnaeus*" (S. Knapp and Q. Wheeler, eds.). The Linnean Society of London, London.

Platnick, N. I. 2012. The poverty of the PhyloCode: a reply to de Queiroz and Donoghue. *Syst. Biol.* 61:360–361.

Platnick, N. I. 2013. The information content of taxon names: a reply to de Queiroz and Donoghue. *Syst. Biol.* 62:175–176.

Pleijel, F. 1999. Phylogenetic taxonomy, a farewell to species, and a revision of *Heteropodarke* (*Hesionidae, Polychaeta, Annelida*). *Syst. Biol.* 48:755–789.

Pleijel, F., and M. Härlin. 2004. Phylogenetic nomenclature is compatible with diverse philosophical perspectives. *Zool. Scr.* 33:587–591.

Pleijel, F., and G. W. Rouse. 2000. A new taxon, *capricornia* (*Hesionidae, Polychaeta*), illustrating the LITU ('least-inclusive taxonomic unit') concept. *Zool. Scr.* 29:157–168.

Pleijel, F., and G. W. Rouse. 2003. Ceci n'est pas une pipe: names, clades and phylogenetic nomenclature. *J. Zool. Syst. Evol. Res.* 41:162–174.

Polaszek, A., and E. O. Wilson. 2005. Sense and stability in animal names. *Trends Ecol. Evol.* 20:421–422.

Rieppel, O. 2006. The PhyloCode: a critical discussion of its theoretical foundation. *Cladistics* 22:186–197.

Roth, B. 1996. Homoplastic loss of dart apparatus, phylogeny of the genera, and a phylogenetic taxonomy of the *Helminthoglyptidae* (*Gastropoda*: *Pulmonata*). *Veliger* 39:18–42.

Rowe, T. 1987. Definition and diagnosis in the phylogenetic system. *Syst. Zool.* 36:208–211.

Rowe, T. 1988. Definition, diagnosis, and origin of *Mammalia*. *J. Vertebr. Paleontol.* 8:241–264.

Rowe, T., and J. Gauthier. 1992. Ancestry, paleontology and definition of the name *Mammalia*. *Syst. Biol.* 41:372–378.

Sangster, G. 2005. A name for the clade formed by owlet-nightjars, swifts and hummingbirds (*Aves*). *Zootaxa* 799:1–6.

Schander, C. 1998a. Types, emendations and names—a reply to Lidén et al. *Taxon* 47:401–406.

Schander, C. 1998b. Mandatory categories and impossible hierarchies—a reply to Sosef. *Taxon* 47:407–410.

Schander, C., and M. Thollesson. 1995. Phylogenetic taxonomy—some comments. *Zool. Scr.* 24:263–268.

Schuh, R. T. 2003. The Linnaean system and its 250-year persistence. *Bot. Rev.* 69:59–78.

Sereno, P. C. 1999. Definitions in phylogenetic taxonomy: critique and rationale. *Syst. Biol.* 48:329–351.

Sereno, P. C. 2005. The logical basis of phylogenetic taxonomy. *Syst. Biol.* 54:595–619.

Smedmark, J. E. E., and T. Eriksson. 2002. Phylogenetic relationships of *Geum* (*Rosaceae*) and relatives inferred from the nrITS and *trnL-trnF* regions. *Syst. Bot.* 27:303–317.

Spangler, R. E. 2003. Taxonomy of *Sarga, Sorghum* and *Vacoparis* (*Poaceae*: *Andropogoneae*). *Aust. Syst. Bot.* 16:279–299.

Stefanovic, S., D. F. Austin, and R. G. Olmstead. 2003. Classification of *Convolvulaceae*: a phylogenetic approach. *Syst. Bot.* 28:791–806.

Stevens, P. F. 2002. Why do we name organisms? Some reminders from the past. *Taxon* 51:11–26.

Stevens, P. F. 2006. An end to all things?—plants and their names. *Aust. Syst. Bot.* 19:115–133.

Stuessy, T. F. 2000. Taxon names are *not* defined. *Taxon* 49:231–233.

Stuessy, T. F. 2001. Taxon names are *still* not defined. *Taxon* 50:185–186.

Sundberg, P., and F. Pleijel. 1994. Phylogenetic classification and the definition of taxon names. *Zool. Scr.* 23:19–25.

Swann, E. C., E. M. Frieders, and D. J. McLaughlin. 1999. *Microbotryum, Kriegeria* and the changing paradigm in basidiomycete classification. *Mycologia* 91:51–66.

Tang, Y.-C., and A.-M. Lu. 2005. Paraphyletic group, PhyloCode and phylogenetic species—the current debate and a preliminary commentary. *Acta Phytotaxon. Sin.* 43:403–419.

Taylor, M. P. 2007. Phylogenetic definitions in the pre-PhyloCode era; implications for naming clades under the PhyloCode. *PaleoBios* 27:1–6.

Taylor, M. P., and D. Naish. 2005. The phylogenetic taxonomy of *Diplodocoidea* (*Dinosauria*: *Sauropoda*). *PaleoBios* 25:1–7.

Wenzel, J. W., K. C. Nixon, and G. Cuccodoro. 2004. Dites non au PhyloCode! *Bull. Soc. Fr. Syst.* 31:19–23.

Wilkinson, M. 2006. Identifying stable reference taxa for phylogenetic nomenclature. *Zool. Scr.* 35:109–112.

Wolfe, A. D., S. L. Datwyler, and C. P. Randle. 2002. A phylogenetic and biogeographic analysis of the *Cheloneae* (*Scrophulariaceae*) based on ITS and *matK* sequence data. *Syst. Bot.* 27:138–148.

Wolsan, M. 2007. Naming species in phylogenetic nomenclature. *Syst. Biol.* 56:1011–1021.

Wyss, A. R., and J. Meng. 1996. Application of phylogenetic taxonomy to poorly resolved crown clades: a stem-modified node-based definition of *Rodentia*. *Syst. Biol.* 45:559–568.

Preamble

1. Biology requires a precise, coherent, international system for naming clades. Scientific names have long been governed by the traditional codes (listed in Preamble item 4), but those codes do not provide a means to give stable, unambiguous names to clades. This code satisfies that need by providing rules for naming clades and describing the nomenclatural principles that form the basis for those rules.

2. This code is applicable to the names of all clades of organisms, whether extant or extinct.

3. This code may be used concurrently with the rank-based codes.

4. Although this code relies on the rank-based codes (i.e., *International Code of Nomenclature for Algae, Fungi, and Plants (Melbourne Code) (ICNAFP*)*, *International Code of Zoological Nomenclature (ICZN)*, *International Code of Nomenclature of Prokaryotes (ICNP)*, *International Code of Virus Classification and Nomenclature (ICVCN)*) to determine the acceptability of preexisting names, it governs the application of those names independently from the rank-based codes.

---

* The abbreviation adopted in the *International Code of Nomenclature for Algae, Fungi, and Plants* itself is *ICN*.

5. This code includes rules, recommendations, notes, and examples. Rules are mandatory in that names contrary to them have no official standing under this code. Recommendations are not mandatory in that names contrary to them cannot be rejected on that basis. Systematists are encouraged to follow them in the interest of promoting nomenclatural uniformity and clarity, but editors and reviewers should not require that they be followed. Notes and examples are intended solely for clarification.

6. This code will take effect on the publication of *Phylonyms: A Companion to the PhyloCode*, and it is not retroactive.

Division I
Principles

1. Reference. The primary purpose of taxon names is to provide a means of referring to taxa, as opposed to indicating their characters, relationships, or membership.
2. Clarity. Taxon names should be unambiguous in their designation of particular taxa. Nomenclatural clarity is achieved through explicit definitions that describe the concept of the taxon designated by the defined name.
3. Uniqueness. To promote clarity, each taxon should have only one accepted name, and each accepted name should refer to only one taxon.
4. Stability. The names of taxa should not change over time. As a corollary, it must be possible to name newly discovered taxa without changing the names of previously discovered taxa.
5. Phylogenetic context. This code is concerned with the naming and subsequent application of the names of phylogenetically conceptualized taxa.
6. Taxonomic freedom. This code does not restrict freedom of opinion with regard to hypotheses about relationships; it only concerns how names are to be applied within the context of any relevant phylogenetic hypothesis.
7. There is no "case law" under this code. Nomenclatural problems are resolved by the Committee on Phylogenetic Nomenclature (CPN) by direct application of the code; previous decisions will be considered, but the CPN is not obligated by precedents set in those decisions.

Division II
Rules

Chapter I

Taxa

ARTICLE 1. CATEGORIES OF TAXA

1.1. The groups of organisms or species considered potential recipients of scientific names are called taxa (singular: taxon). The only taxa whose names are governed by this code are clades. However, species, whose names are governed by the rank-based codes, are frequently used to define clade names in this code.

ARTICLE 2. CLADES

2.1. In this code, a clade is an ancestor (an organism, population, or species) and all of its descendants.

Note 2.1.1. Every individual organism (on Earth) belongs to at least one clade (i.e., the clade comprising all extant and extinct organisms, assuming that they share a single origin). Each organism also belongs to a number of nested clades (though the ancestor of the clade comprising all life—again assuming a single origin—does not belong to any other clade).

Note 2.1.2. It is not necessary that all clades be named.

Note 2.1.3. Clades are often either nested or mutually exclusive; however, phenomena such as speciation via hybridization, species fusion, and symbiogenesis can result in clades that are partially overlapping (Fig. 1).

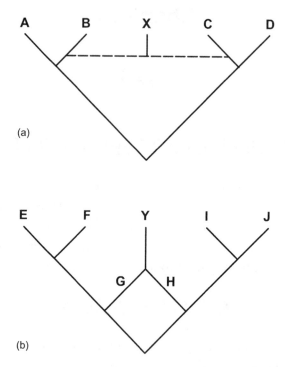

(a)

(b)

Fig. 1. Speciation via hybridization (a) and species fusion (b) can result in clades that are partially overlapping. In (a), the origin of species X via hybridization (represented by the dashed line) between members of species B and C results in partial overlap between the largest clade containing A but not D (or the smallest clade containing both A and B), which is composed of A, B, and X, and the largest clade containing D but not A (or the smallest clade containing C and D), which is composed of C, D, and X, in that X is part of both clades. In (b), fusion of species G and H to form species Y (with the two parent species disappearing in the process) results in partial overlap between the largest clade containing E but not J (or the smallest clade containing both E and G), which is composed of E, F, G, and Y, and the largest clade containing J but not E (or the smallest clade containing both H and J), which is composed of H, I, J, and Y, in that Y is part of both clades.

2.2. In this code, a clade is called a crown clade if it originates in the most recent common ancestor of two or more extant species or organisms. A clade is called a total clade if it is composed of a crown clade and all species or organisms that share a more recent common ancestor with that crown clade than with any extant species or organisms that are not members of that crown clade. This code governs all formal clade names, many of which designate neither crown nor total clades.

ARTICLE 3. HIERARCHY AND RANK

3.1. The system of nomenclature described in this code is independent of categorical rank (e.g., genus, family, etc.). Although clades are hierarchically related, and therefore intrinsically ranked in the sense that some are more inclusive than others, assignment of categorical ranks is not part of the formal naming process and has no bearing on the spelling or application of clade names.

Example 1. If the name *Iguanidae* were defined as referring to a clade originally ranked as a family, and if that clade were later ranked as a subfamily and (at the same time) a larger clade ranked as a family, the reference of the name *Iguanidae* would not change to the larger clade, nor would the spelling of that name change (i.e., to *Iguaninae*) to reflect the new rank of the clade to which it refers.

Note 3.1.1. This code does not prohibit, discourage, encourage, or require the use of categorical ranks.

3.2. The concepts of synonymy, homonymy, and precedence adopted in this code (see Arts. 12–14) are, in contrast to the rank-based codes, independent of categorical rank.

Chapter II

Publication

ARTICLE 4. PUBLICATION REQUIREMENTS

4.1. The provisions of this article apply not only to the publication of names, but also to the publication of any nomenclatural act (e.g., a proposal to conserve a name).

4.2. Publication, under this code, is defined as the distribution of peer-reviewed works consisting of: (1) printed text with or without images, which, unless also published electronically, must be distributed to libraries or scientific institutions associated with libraries in at least five countries on three continents, so that the work is accessible as a permanent public record to the scientific community; and/or (2) electronic text with or without images or sound in Portable Document Format (PDF) in an online publication (however, not just in supplementary material; see Note 7.2.2); in both cases with an International Standard Serial Number (ISSN) or an International Standard Book Number (ISBN).

Note 4.2.1. If a manuscript is released electronically in stages (e.g., accepted manuscript, uncorrected proofs, corrected proofs without pagination, and final version with pagination), the final version with pagination is to be considered the "publication" for the purposes of this code.

Note 4.2.2. If an entire book is not peer-reviewed or a periodical is not consistently peer-reviewed, the article or chapter in which a name or nomenclatural act appears must be peerreviewed in order to qualify as published.

Note 4.2.3. Approval of a work by a thesis or dissertation committee does not constitute peer review.

Note 4.2.4. The distribution before the publication date of *Phylonyms: A Companion to the PhyloCode* of any material (printed or electronic) does not constitute publication (see Art. 7.1).

4.3. For the purpose of Article 4.2, "online" is defined as accessible electronically via the World Wide Web.

4.4. Should Portable Document Format (PDF) be succeeded, a successor international standard format approved and communicated by the Committee on Phylogenetic Nomenclature would be acceptable.

Recommendation 4.4A. Publication electronically in Portable Document Format (PDF) should comply with the PDF/A archival standard (ISO 19005).

Recommendation 4.4B. Authors of electronic material should give preference to publications that are archived and curated in trusted online digital repositories, e.g., an ISO-certified repository. Digital repositories ideally should be in more than one country, preferably on different continents.

4.5. An electronic publication (see Note 4.2.1) must not be altered after it is published. Any such alterations are not themselves considered published. Corrections or revisions must be issued separately to be considered published.

4.6. The following do not qualify as publication: (a) dissemination of text or images solely through storage media (such as CDs, diskettes, film, microfilm, and microfiche) that require a special device to read; (b) theses and dissertations; (c) abstracts of articles, papers, posters, texts of lectures, and similar material presented at meetings, symposia, colloquia, or congresses, even if the abstract is printed in a peer-reviewed journal; (d) the placing of texts or images in collections or exhibits, for example, on labels (including specimen labels, even if printed) or information sheets; (e) the reproduction of hand-written material in facsimile, for example, by photocopy; (f) patents and patent applications; (g) newspapers and periodicals intended mainly for people who are not professional scientists, abstracting journals, trade catalogues, and seed exchange lists; (h) anonymous works. See also Article 7.3.

ARTICLE 5. PUBLICATION DATE

5.1. The publication date for names established under this code is the date on which a publication, as defined in Article 4 (see especially Note 4.2.1), first becomes available either in print or online. In the absence of proof establishing some other date, the one appearing in the publication itself must be accepted as correct.

5.2. If the date appearing in the publication specifies the month but not the day, the last day of that month is to be adopted as the publication date.

5.3. If the date appearing in the publication specifies the year but not the month, the last day of that year is to be adopted as the publication date.

5.4. When separates are issued in advance of the work (periodical or book) that contains them, the date of the separate constitutes the date of publication, unless there is evidence that it is erroneous.

Chapter III

Names

ARTICLE 6. CATEGORIES OF NAMES

6.1. Established clade names are those that are published in accordance with Article 7 of this code. Unless a clade name is established, it has no status under this code.

Recommendation 6.1A. In order to distinguish scientific names from other (e.g., vernacular) names, all scientific names should be italicized when they appear in print.

Note 6.1A.1. Italicizing all scientific names is consistent with the 2018 edition of the *ICNAFP* but not with the 1999 edition of the *ICZN*.

Recommendation 6.1B. In order to indicate which clade names are established under this code and therefore have explicit phylogenetic definitions (and whose endings are not reflective of rank), it may be desirable to distinguish these names from supraspecific names governed by the rank-based codes, particularly when both are used in the same publication.

Example 1. The letter "P" (bracketed or in superscript) might be used to designate clade names governed by this code, and the letter "R" to designate names governed by the rank-based codes. Using this convention, the name "*Ajugoideae*[R]" would apply to a plant subfamily, which may or may not be a clade, whereas "*Teucrioideae*[P]" would apply to a clade, which may or may not be a subfamily.

Example 2. If the name *Teucrioideae* applied to both a clade (this code) and a subfamily (*ICNAFP*), they could be distinguished as "clade *Teucrioideae*" versus "subfamily *Teucrioideae*."

6.2. Preexisting names are scientific names that prior to their establishment under this code were either: (a) "legitimate" (*ICNAFP*, *ICNP*), "potentially valid" (*ICZN*), or "valid" (*ICVCN*); or (b) in use but whose application to taxa is not governed by any code (e.g., unranked names and zoological names ranked above superfamily). In addition, scientific names governed by the *ICNAFP* that are in current or recent use but have never been published with a Latin description or diagnosis (and therefore violate *ICNAFP* Article 39 if published between 1935 and 2011) are considered under this code to be preexisting names, provided that they have been published with a description or diagnosis in some other language and otherwise qualify as legitimate names under the *ICNAFP*.

Note 6.2.1. Names that were phylogenetically defined in publications (Art. 4) prior to the starting date of this code (Art. 7.1) and are not "legitimate" (*ICNAFP*, *ICNP*), "potentially valid" (*ICZN*), or "valid" (*ICVCN*) are considered to be preexisting names after the starting date of this code. They fall under Article 6.2(b) because they are in use but were not governed by any code at the time they were published.

6.3. Converted names are preexisting names that have been established as clade names in accordance with the rules of this code.

6.4. An acceptable clade name is one that is in accordance with the rules of this code; that is, it is both (a) established and (b) not a non-conserved (Art. 15) later homonym.

6.5. The accepted name of a clade is the name that must be adopted for it under this code. It must (1) be established (Art. 7), (2) have precedence (Arts. 12–15) over alternative uses of the same name (homonyms) and alternative names for the same clade (synonyms), and (3) not be rendered inapplicable by a qualifying clause in the context of a particular phylogenetic hypothesis (Art. 11.12).

6.6. Once a clade name has been established, its status as an acceptable and/or accepted name is not affected by inaccurate or misleading connotations; thus, a name is not to be rejected because of a claim that it denotes a character, distribution, or relationship not possessed by the clade.

ARTICLE 7. GENERAL REQUIREMENTS FOR ESTABLISHMENT

7.1. Establishment of a name can only occur through publication in *Phylonyms: A Companion to the PhyloCode*, or in another work after the publication date of *Phylonyms*, the starting date for this code.

7.2. In order to be established, a name of a clade must: (a) be published as provided for by Article 4; (b) be adopted by the author(s), not merely proposed for the sake of argument or on the condition that the group concerned will be accepted in the future; (c) apply to a clade that either appears on the reference phylogeny or is delimited by the cited apomorphy(-ies) (see Art. 9.13); (d) comply with the provisions of Articles 7 and 9–11; (e) be registered as provided for in Article 8, and the registration number be cited in the protologue; and (f) comply with the provisions of Article 17.

Note 7.2.1. The protologue is everything associated with a name when it was first established (this code), validly published (*ICNAFP*, *ICNP*), or made available (*ICZN*), for example, description, diagnosis, phylogenetic definition, registration number, designation of type, illustrations, references, synonymy, geographical data, specimen citations, and discussion.

Note 7.2.2. Material contained only in an electronic supplement to a printed or online journal is not published according to Article 4. Therefore, the following portions of the protologue must not be confined to an electronic supplement: (a) clade name to be established; (b) designation of clade name as new or converted (Art. 9.2); (c) phylogenetic definition (Arts. 9.3, 9.4); (d) reference phylogeny or statement about the distribution of apomorphies supporting the existence of the clade (Art. 9.13); (e) hypothesized composition of the clade (Art. 9.14); (f) for converted names, bibliographic citations (Art. 9.16) demonstrating prior application of the name to a taxon approximating the clade for which it is being established (Art. 9.15a) and authorship of the preexisting name (Art. 9.15b); (g) registration number (Art. 7.2e); (h) when appropriate, the rationale for selection of the name (e.g., Recs. 10.1A, 10.1B).

7.3. When a publication contains a statement to the effect that names or nomenclatural acts in it are not to be considered for nomenclatural purposes, names that it may contain are considered as not established.

ARTICLE 8. REGISTRATION

8.1. In order for a name to be established under this code, the name and other required information must be submitted to the registration database for phylogenetically defined names (see Art. 22.2). A name may be submitted to the database prior to acceptance for publication, but it is given only a temporary registration number at that time. The registration number will become permanent after the author notifies the database that the paper or book in which the name will appear has been published, provides a full reference to the publication, and confirms that the definition in the database is identical to that in the publication.

Note 8.1.1. Specification of the data that are required for registration can be obtained via the Internet or directly from the database administrator. The registration procedure and a provisional list of required data are found in Appendix A.

Recommendation 8.1A. A name should not be submitted to the registration database more than one month before it is submitted for publication, to prevent names from being reserved indefinitely in anticipation of possible publication.

Recommendation 8.1B. Registration of a name whose spelling or definition is identical to one that already exists in the database should generally be avoided (but see Recs. 8B, 8C). However, such names are not treated by this code as homonyms or synonyms until published.

8.2. At the submitter's request, a name or definition that he or she proposed can be changed or removed from the registration database if it is not yet published.

Recommendation 8.2A. The submitter of an unpublished registered name or definition who decides to change it or not to publish it should notify the database administrator promptly.

8.3. If the registered definition of a name disagrees with the definition in the protologue or the name is defined more than one way in the protologue, the author should determine which is correct and notify the registration database administrator promptly.

Note 8.3.1. If the author notifies the database administrator that the registered definition is incorrect, the administrator will correct the database and insert a note that the change was made. If one or more definitions in the protologue are incorrect, the administrator will annotate the database to alert users that this is the case.

8.4. If the registered definition of a name disagrees with the definition in the protologue or the name is defined more than one way in the protologue, and the author is no longer alive or is otherwise unable to determine which definition is correct, the following guidelines are to be used: If it is clear that the differences between the definitions are due to typographical errors, the definition that lacks typographical errors is treated as correct. If it is not clear that the differences between the definitions are due to typographical errors, the definition immediately associated with the designation "new clade name," "converted clade name," etc. is treated as correct. If two or more definitions are equally closely associated with the designation "new clade name" or "converted clade name," the decision as to which is considered correct is to be based on an interpretation of the author's intent. Such decisions regarding the correct definition of a name, if made by anyone other than the author, must be published (Art. 4) before the registration database administrator is notified (see Rec. 8A). Once published, such decisions can be reversed only by the CPN.

Note 8.4.1. If the author of a published correction notifies the database administrator that the registered definition is incorrect, the administrator will correct the database and insert a note that the change was made. If one or more definitions in the protologue are incorrect, the administrator will annotate the database to alert users that this is the case.

8.5. If the registered definition of a name and the definition in the protologue agree but contain a typographical error, the author may publish a correction. If the author is no longer alive or is otherwise unable to correct the error, any person may publish a correction (see Rec. 8A).

Note 8.5.1. After the registration database administrator is notified, the definition will be corrected in the database, and a note will be added stating that the change was made.

Note 8.5.2. A correction slip inserted in the original publication does not qualify as a published correction. Publication of corrections must satisfy the requirements of Article 4.

8.6. Accidental errors in a definition that appear in print subsequent to establishment are not to be treated as new definitions (i.e., establishment of homonyms) but as incorrect statements of the established definition. The same is true of unjustified corrections (i.e., any correction that does not fall under Arts. 8.3–8.5).

Recommendation 8A. The person making corrections of the sort covered by Articles 8.4 and 8.5 should notify the database administrator promptly after publishing it.

Recommendation 8B. If a name or definition has been registered, but there is no indication in the registration database whether it was ever published, the name or definition should not be published by another person who has not first attempted to determine whether it was ever published. If bibliographic databases fail to resolve the question, a serious effort should be made to contact the person who registered the name or definition. (Contact information submitted with the name and maintained in the database may facilitate this.)

Recommendation 8C. If a serious but unsuccessful attempt has been made to determine whether a registered name was ever published, and the name is new (not based on a preexisting name), it is better to choose a different name, rather than use the same name and risk creating a homonym. If, in the same situation, the registered name is based on a preexisting name, it is better to publish a definition of this name, even at the risk of creating a homonym, rather than choose another, less appropriate name. This is particularly true if the registered name is widely used.

Chapter IV

Establishment of Clade Names

ARTICLE 9. GENERAL REQUIREMENTS AND PHYLOGENETIC DEFINITIONS

9.1. The names of clades may be established through conversion of preexisting names or introduction of new names.

9.2. In order to be established, converted clade names must be clearly identified as such in the protologue by the designation "converted clade name" or "*nomen cladi conversum.*" New clade names must be identified as such by the designation "new clade name" or "*nomen cladi novum.*"

9.3. In order to be established, a clade name must be provided with a phylogenetic definition, written in English or Latin, linking it explicitly with a particular clade. The name applies to whatever clade fits the definition.

Note 9.3.1. The application of a phylogenetic definition, and thus also of a phylogenetically defined clade name, requires a hypothesized phylogeny. To accommodate phenomena such as speciation via hybridization, species fusion, and symbiogenesis (see Note 2.1.3), the hypothesized phylogeny that serves as the context for the application of a phylogenetically defined name need not be strictly diverging.

9.4. A phylogenetic definition is a statement that explicitly identifies a particular clade as the referent of a taxon name. Different categories of acceptable phylogenetic definitions include, but are not limited to, those described in Articles 9.5–9.7 and 9.9–9.10. Articles 9.5–9.7 describe general categories; Articles 9.9 and 9.10 describe categories designed for naming crown clades and total clades, respectively. Qualifying clauses are described in Article 11.12.

Note 9.4.1. The following conventions are adopted for abbreviated definitions: max = the largest; min = the smallest; ∇ = clade; () = containing (but see Note 10.5.1); [] = as exhibited by; apo = characterized by apomorphy [followed by the name or description of the apomorphy]; & = and; ∨ = or; ~ = but not; A, B, C, etc. = species or specimens used as internal specifiers (see Art. 11.2); Z, Y, X, etc. = species or specimens used as external specifiers (see Art. 11.2); M = an apomorphy used as an internal specifier.

Recommendation 9.4A. Because poorly chosen wordings of phylogenetic definitions can lead to undesirable consequences (i.e., the application of the name in a way that contradicts the author's intent), the wordings provided in Articles 9.5–9.7 and 9.9–9.10 should generally be used for the corresponding kinds of definitions. If an alternative wording is used, it should be accompanied by the standard abbreviation (as provided in Arts. 9.5–9.7, Arts. 9.9–9.10 and Note 9.4.1) to clarify the intent of the author in case the alternative wording is ambiguous or confusing. If the definition in words and its abbreviated form appear to be in conflict, the latter should be weighted most heavily in interpreting the author's intent. This recommendation does not preclude the use of other kinds of definitions that are not addressed in Articles 9.5–9.7 and 9.9–9.10.

9.5. A minimum-clade definition (formerly known as a node-based definition*) associates a name with the smallest clade that contains two or more internal specifiers. Such a definition may take the form "the clade originating in the most recent common ancestor of A and B" or "the smallest clade containing A and B", where A and B are internal specifiers (Art. 11.2). A minimum-clade definition may be abbreviated "min ∇ (A & B)". Additional internal specifiers (e.g., C & D & E, etc.) may be used as needed (e.g., if the basal relationships within the clade are poorly supported). For defining the names of crown clades using minimum-clade definitions, see Article 9.9.

Note 9.5.1. A directly-specified-ancestor definition is a special case of the minimum-clade definition in which the ancestor in which the clade originated is specified directly rather than indirectly through its descendants. A directly-specified-ancestor definition may take the form "the clade originating in A" or "the smallest clade containing A", where A is an internal specifier (Art. 11.2). A directly-specified-ancestor definition may be abbreviated "min ∇ (A)". For defining the names of crown clades using directly-specified-ancestor definitions, see Note 9.9.2.

Note 9.5.2. Provided that the internal specifiers have a common ancestor, a minimum-clade definition as described in Article 9.5 necessarily identifies a clade; there can be disagreements about the composition of the clade when the definition is applied in the context of different phylogenetic hypotheses, but not about its existence. It is possible to formulate a minimum-clade definition according to which the defined name does not apply to any clade under particular phylogenetic hypotheses through the use of a qualifying clause (see Art. 11.12) or external specifiers (see Art. 11.13, Ex. 1).

---

* Although both sets of terms refer to the same definitional forms (wordings) described in Articles 9.5, 9.6, and 9.9, the definitional concepts to which they refer are not necessarily identical (and thus the terms are not strictly equivalent). For example, a minimum-clade definition is not necessarily a node-based definition in that the most recent common ancestor of A and B need not be interpreted as corresponding to a node representing a lineage at the instant of a splitting event; it could instead be interpreted as corresponding to a branch representing an entire ancestral species.

9.6. A maximum-clade definition (formerly known as a branch-based or a stem-based definition*) associates a name with the largest clade that contains one or more internal specifiers but does not contain one or more external specifiers. Such a definition may take the form "the clade consisting of A and all organisms or species that share a more recent common ancestor with A than with Z" or "the clade originating in the earliest ancestor of A that is not an ancestor of Z" or "the largest clade containing A but not Z", where A is an internal specifier (Art. 11.2) and Z is an external specifier (Art. 11.2). A maximum-clade definition may be abbreviated "max ∇ (A ⁓ Z)". Additional external specifiers (e.g., Y & X & W, etc.) may be used as needed (e.g., if the sister group of the named clade is uncertain). For defining the names of crown clades using maximum-clade definitions, see Article 9.9; for defining the names of total clades using maximum-clade definitions, see Article 9.10.

Note 9.6.1. Provided that the internal and external specifiers have a common ancestor, a maximum-clade definition as described in Article 9.6 necessarily identifies a clade; there can be disagreements about the composition of the clade when the definition is applied in the context of different phylogenetic hypotheses, but not about its existence. It is possible to formulate a maximum-clade definition according to which the defined name does not apply to any clade under particular phylogenetic hypotheses through the use of a qualifying clause (see Art. 11.12) or multiple internal specifiers (see Art. 11.13, Ex. 2).

Note 9.6.2. It is important to use the appropriate operator, "and" ("&") versus "or" ("∇"), in definitions employing multiple external specifiers (only the "and" operator would normally be used in definitions employing multiple internal specifiers). The "and" operator is to be used when the intent is to exclude jointly all of the external specifiers from the named clade. For example, it would be appropriate to use "and" when using a maximum-clade definition with

---

* See footnote to Article 9.5.

multiple external specifiers to deal with uncertainty regarding the sister group of the named clade—that is, to exclude jointly every taxon that is a potential sister group. By contrast, the "or" operator is to be used when the intent is to exclude, whether individually or jointly, any one (or more) of the external specifiers from the named clade. For example, it would be appropriate to use "or" when using a minimum-clade definition with multiple external specifiers, including those used in qualifying clauses, to render the defined name inapplicable in the context of phylogenetic hypotheses in which any one (or more) of the external specifiers is more closely related to some of the internal specifiers than those internal specifiers are to other internal specifiers (see Art. 11.12, Ex. 1).

9.7. An apomorphy-based definition associates a name with a clade originating in the first ancestor to evolve a specified apomorphy that was inherited by one or more internal specifiers. Such a definition may take the form "the clade originating in the ancestor in which apomorphy M, as inherited by A, originated" or "the clade for which M, as inherited by A, is an apomorphy" or "the clade characterized by apomorphy M as inherited by A", where A is an internal specifier species or specimen and M is a specifier apomorphy (Arts. 11.1–11.2). An apomorphy-based definition may be abbreviated "∇ apo M[A]". Additional internal specifiers may be used as needed (e.g., if one intends for the name not to apply to any clade in cases in which the specified apomorphy is shared by those specifiers as the result of homoplasy). For defining the names of crown clades using apomorphy-based definitions, see Article 9.9.

Note 9.7.1. An apomorphy-based definition as described in Article 9.7 necessarily identifies a clade provided that there is only one internal specifier; there can be disagreements about the composition of the clade when the definition is applied in the context of different phylogenetic hypotheses, but not about its existence. It is possible to formulate an apomorphy-based definition according to which the defined name does not apply to any clade under particular phylogenetic hypotheses through the use of a qualifying clause (see Art. 11.12) or multiple internal specifiers (see Art. 11.13, Ex. 3).

Recommendation 9.7A. If an apomorphy-based definition is used, or if an apomorphy is cited in a qualifying clause (see Art. 11.12, Ex. 2), the apomorphy should be described or illustrated in sufficient detail that users of the definition will understand the author's intent.

Recommendation 9.7B. If an apomorphy-based definition is used, and if the apomorphy is a complex character that could have evolved in a stepwise fashion, then the author should identify in the protologue which aspect(s) of that apomorphy must be present in order for an organism to be considered to belong to the clade whose name is defined by that apomorphy.

9.8. If the author of an apomorphy-based definition based on a complex apomorphy did not identify in the protologue which aspect(s) of that apomorphy must be present in order for an organism to be considered to belong to the clade whose name is defined by that apomorphy (Rec. 9.7B), or if an aspect that the author did identify is later found to be a complex apomorphy itself, then subsequent authors are to interpret the definition as applying to the clade characterized by the presence of all of the components of the complex apomorphy described by the author of the definition (see Rec. 9.7A) or present in the taxa or specimens that the author of the definition considered to possess that apomorphy. Similarly, if multiple apomorphies are used in an apomorphy-based definition, subsequent authors are to interpret the definition as applying to the clade characterized by the presence of all of those apomorphies.

9.9. A crown-clade definition is a phylogenetic definition that necessarily identifies a crown clade (Art. 2.2) as the referent of a taxon name.

■ A minimum-clade definition (Art. 9.5) is a crown-clade definition if all of the internal specifiers (Art. 11.2) are extant, or if the definition is explicitly stated as applying to the name of a crown clade. A minimum-crown-clade definition may thus take the form "the crown clade originating in the most recent common ancestor of A and B" or "the smallest crown clade containing A and B", where A and B are internal specifiers. A minimum-crown-clade definition may be abbreviated "min crown ∇ (A & B)". Additional internal specifiers (e.g., C & D & E, etc.) may be used as needed (e.g., if the basal relationships within the clade are poorly supported). If this kind of definition is used and "extant" is intended to mean anything other than extant on the publication date of the definition (thus affecting the concept of a crown clade; see Art. 2.2), the author should specify the intended meaning (within the restrictions described in Art. 9.11)—e.g., the internal specifiers were extant (and thus the clade was a crown clade) at a particular time in human history.

Note 9.9.1. Minimum-crown-clade definitions can be either implicit (if all of the internal specifiers are extant but application to the name of a crown clade is not expressly stated) or explicit (if application to the name of a crown clade is expressly stated).

Note 9.9.2. If a directly-specified-ancestor definition (see Note 9.5.1) is used and the single internal specifier is an extant species, then the named clade is a crown clade (see Art. 2.2) consisting of that species and any descendants it might have. Such a definition may take the form "the crown clade originating in A" or "the smallest crown clade containing A", where A is an extant internal specifier (Art. 11.2). It may be abbreviated "min crown ∇ (A)". This kind of definition may be useful in defining the names of crown clades comprising single extant species.

■ A maximum-clade definition (Art. 9.6) is a crown-clade definition if (1) at least one of the (explicitly mentioned) internal specifiers (Art. 11.2) is extant and (2a) the word "extant" is included before "organisms" under the first wording (Art. 9.6) or (2b) the word "crown" is included before "clade" under the third wording (Art. 9.6). A maximum-crown-clade definition (formerly known as a branch-modified or a stem-modified node-based definition*) may thus take the form "the crown clade originating in the most recent common ancestor of A and all extant organisms or species that share a more recent common ancestor with A than with Z" or "the largest crown clade containing A but not Z", where A is an extant internal specifier and Z is an external specifier (Art. 11.2). A maximum-crown-clade definition may be abbreviated "max crown ∇ (A ~ Z)". Additional internal specifiers (e.g., C & D & E, etc.) and external specifiers (e.g., Y & X & W, etc.) may be used as needed (e.g., if the extant outgroup relationships of the named clade are poorly supported in the case of external specifiers, or if the author intends for the name not to apply to any clade in the context of particular phylogenetic hypotheses in the case of internal specifiers, as described in Article 11.13, Ex. 2; but see Note 9.6.2). If this kind of definition is used and "extant" is intended to mean anything other than extant on the publication date of the definition (thus affecting the concept of a crown clade; see Art. 2.2), the author should specify the intended meaning (within the restrictions described in Art. 9.11)—e.g., the internal specifiers were extant (and thus the clade was a crown clade) at a particular time in human history.

- An apomorphy-based definition is not to be used to define the name of a crown clade, as that practice requires certainty that the defining apomorphy and the crown clade originated in the same ancestor. Nonetheless, a minimum-clade definition can be modified by the use of an apomorphy to define the name of a crown clade. An apomorphy-modified

---

* See footnote to Article 9.5.

crown-clade definition (formerly known as an apomorphy-modified node-based definition*) may thus take the form "the crown clade originating in the most recent common ancestor of A and all extant organisms or species that inherited M synapomorphic with that in A" or "the crown clade for which M, as inherited by A, is an apomorphy relative to other crown clades," or "the crown clade characterized by apomorphy M (relative to other crown clades) as inherited by A", where (in all three wordings) A is an extant specifier species or specimen and M is a specifier apomorphy (Arts. 11.1–11.2). An apomorphy-modified-crown-clade definition may be abbreviated "crown ∇ apo M[A]". If this kind of definition is used and "extant" is intended to mean anything other than extant on the publication date of the definition (thus affecting the concept of a crown clade; see Art. 2.2), the author must indicate explicitly or implicitly the intended meaning (within the restrictions described in Art. 9.11)—e.g., that the internal specifiers were extant (and thus the clade was a crown clade) at a particular time in human history.

Note 9.9.3. If some or all of the internal specifiers are extinct in a minimum-clade definition or if all of the internal specifiers are extinct in a maximum-clade definition or an apomorphy-based definition, and if the name is not explicitly stated as applying to the name of a crown clade, then the defined name may nevertheless apply to a crown clade in the context of a particular phylogenetic hypothesis. However, it is not considered to be a crown-clade definition because the defined name does not necessarily apply to a crown clade in the context of all relevant phylogenetic hypotheses.

Recommendation 9.9A. When a minimum-clade definition is intended to define the name of a crown clade, application to a crown clade should be stated explicitly.

---

* See footnote to Article 9.5.

9.10. A total-clade definition is a phylogenetic definition that necessarily identifies a total clade (Art. 2.2) as the referent of a taxon name.

- A minimum-clade definition is not to be used to define the name of a total clade, as that practice would require certainty that the internal specifiers represent both branches of the earliest divergence within the total clade.
- A maximum-clade definition is a total-clade definition if at least one of the internal specifiers (Art. 11.2) and all of the external specifiers (Art. 11.2) are extant, or if the definition is explicitly stated as applying to the name of a total clade. A maximum-total-clade definition may thus take the form "the total clade consisting of A and all organisms or species that share a more recent common ancestor with A than with Z" or "the total clade originating in the earliest ancestor of A that is not an ancestor of Z" or "the largest total clade containing A but not Z", where A is an internal specifier (Art. 11.2) and Z is an external specifier (Art. 11.2). A maximum-total-clade definition may be abbreviated "max total ∇ (A ~ Z)". Additional internal specifiers (e.g., B & C & D, etc.) and external specifiers (e.g., Y & X & W, etc.) may be used as needed (e.g., if the outgroup relationships of the named clade are poorly supported in the case of external specifiers, or if the author intends for the name not to apply to any clade in the context of particular phylogenetic hypotheses in the case of internal specifiers, as described in Art. 11.13, Ex. 2). If this kind of definition is used and "extant" is intended to mean anything other than extant on the publication date of the definition (thus affecting the concept of a total clade; see Art. 2.2), the author must indicate explicitly or implicitly the meaning of "extant" (within the restrictions described

in Art. 9.11)—e.g., that the relevant specifiers were extant (and thus the clade was a total clade) at a particular time in human history.

■ An apomorphy-based definition is not to be used to define the name of a total clade, as that practice requires certainty that the defining apomorphy and the total clade originated in the same ancestor.

■ Total-clade definitions may also be formulated through reference to their corresponding crown clades. The following are examples of crown-based total-clade definitions: (1) "the total clade of the crown clade [name of crown clade]" (abbreviated "total ∇ of X", where X is the name of a crown clade); for the use of this type of definition for panclade names, see Article 10.5; (2) "the total clade of the smallest crown clade containing A and B", where A and B are extant internal specifiers (abbreviated "total ∇ of min crown ∇ (A & B)"); (3) "the total clade of the largest crown clade containing A but not Z", where A is an extant internal specifier and Z is a specifier (not necessarily extant) that is external to the crown clade (abbreviated "total ∇ of max crown ∇ (A ∼ Z)"; and (4) "the total clade of the crown clade for which M, as inherited by A, is an apomorphy relative to other crown clades", where A is an extant internal specifier organism or species and M is an apomorphy that occurs in it (abbreviated "total ∇ of crown ∇ apo M [A]"). For alternative ways of wording these definitions, see Article 9.9.

Note 9.10.1. Maximum-total-clade definitions can be either implicit (if at least one of the internal specifiers and all of the external specifiers are extant but application to the name of a total clade is not expressly stated) or explicit (if application to the name of a total clade is expressly stated).

Note 9.10.2. Crown-based total-clade definitions with formulation 1 (including those of panclade names; see Art. 10.5) differ from all other definitions described in Articles 9.5–9.7 and 9.9–9.10 in not using any explicit specifiers, which instead are implicit. The internal specifiers in such definitions are those of the crown-clade name used in the definition of the total-clade name. The external specifiers are all extant species or organisms that are not members of that crown clade.

Recommendation 9.10A. When a maximum-clade definition is intended to define the name of a total clade, application to a total clade should be stated explicitly.

9.11. It is permissible to establish a name with a crown-clade definition using an internal specifier that is not extant on the publication date under the following conditions: If that internal specifier is a species, either the specifier must have been extant as of 1500 CE or there must be specimens of the specifier species in existence that were collected when that species was extant. If that internal specifier is a specimen, the organism must either have died in or after 1500 CE or have been alive when it was collected. If an author of a definition intends to define "extant" as anything other than extant on the publication date of the definition, the intended meaning must be indicated explicitly or implicitly in the protologue.

Example 1. If a name were defined as applying to the smallest crown clade containing *Alca torda* and *Pinguinus impennis*†, use of *Pinguinus impennis*† (which became extinct during the nineteenth century) as an internal specifier in the definition would implicitly indicate that the author's concept of that crown clade is based on species that were extant during the nineteenth century, even if currently extinct. By contrast, if the name were defined as applying to the largest crown clade containing *Alca torda* but not *Alle alle* (which are both extant), but the author(s) intended the name to apply to a crown clade that includes *Pinguinus impennis*†, they would have to state explicitly that their concept of the named crown clade is based on species that were extant during the nineteenth century.

9.12. If the author of a crown-clade definition (Art. 9.9) did not indicate explicitly or implicitly the meaning of "extant" or "crown clade" (see Art. 9.11), then subsequent authors are to interpret that definition as referring to organisms or species that were extant on its publication date (Art. 5).

9.13. In order for a clade name to be established, the protologue must include citation of a published reference phylogeny or an explicit statement about the distribution of one or more putative apomorphies supporting the existence of the clade being named. A reference phylogeny is a phylogenetic hypothesis that provides a context for applying a clade name by means of its phylogenetic definition. See Article 11.11 concerning the inclusion of specifiers in the reference phylogeny.

Note 9.13.1. A reference phylogeny is not part of the definition and does not prevent the name from being applied in the context of alternative phylogenies.

Note 9.13.2. The reference phylogeny may be published in the same work in which the name is being established, or a previously published phylogeny may be cited.

Recommendation 9.13A. A reference phylogeny should be derived via an explicit, reproducible analysis.

Recommendation 9.13B. If more than one reference phylogeny is cited in the protologue, one of them, and ideally a single figure or tree, should be designated as the primary reference phylogeny.

9.14. In order for a clade name to be established, the protologue must include a statement about the hypothesized composition of the clade (e.g., a list of included subclades, species, or specimens or reference to such a list). Any specimen citation must include the name of a species or clade (less inclusive than the one whose composition is being described) to which the specimen can be referred, unless the clade whose composition is being described does not contain any named species or clades.

9.15. In order for conversion to be effected, the preexisting name that is being converted to a phylogenetically defined clade name must be indicated. Direct and unambiguous bibliographic citations (as detailed in Art. 9.16) must be provided demonstrating (a) prior application of the name to a taxon approximating the clade for which it is being established (or to a paraphyletic group originating in the same ancestor; see Art. 10.1) and (b) authorship of the preexisting name (but see Rec. 9.15A) for the purpose of attribution (see Arts. 19, 20). In some cases, a single bibliographic citation will serve both purposes, but two different publications will have to be cited if the composition associated with the name by the original author differs substantially from that of the clade for which the converted name is being established (see Art. 19.1).

Note 9.15.1. Errors in the bibliographic citation for a preexisting name should be corrected by subsequent authors, but they do not invalidate the establishment of the corresponding converted name.

Note 9.15.2. Demonstrating "prior application of the name to a taxon approximating the clade for which it is being established" does not necessarily require a modern phylogenetic analysis, and it does not require that the author of the prior application conceptualized the taxon as a clade. Application of a name in an earlier publication to a taxon approximating the clade for which it is being converted can be demonstrated based on information in that work—e.g., a list of subordinate taxa that are broadly consistent with, though not necessarily identical to, the composition of that clade, a description including diagnostic characters that we now understand to be apomorphies of that clade, or statements and diagrams about phylogenetic relationships. When composition is used to assess the prior application of a name, the historical inclusion of taxa that are no longer considered to belong to the clade in question, or the historical exclusion of taxa that are now considered to belong to this clade, does not necessarily disqualify it as a preexisting name for the clade, provided that its application to that clade approximates traditional use to the degree that it is consistent with the contemporary concept of monophyly.

Example 1. Olmstead and Judd in *Phylonyms* applied the preexisting name *Lamianae* to the smallest clade containing *Gentianales*, *Solanales*, *Lamiales*, *Boraginaceae*, and *Vahliaceae*. The name *Lamianae* was first used by Takhtajan, who applied it to a taxon that included *Gentianales*, *Solanales*, *Lamiales*, and *Boraginaceae* (though in some cases under different names) as well as some smaller taxa (e.g., *Dipsacales*, *Polemoniaceae*) that render his *Lamianae* polyphyletic in the context of currently accepted phylogenies. Takhtajan's inclusion of these taxa that are now considered to lie well outside the clade in question, and his omission of *Vahliaceae*, which is now thought to be part of that clade, do not disqualify *Lamianae* as a preexisting name for that clade.

Recommendation 9.15A. If possible, the bibliographic citation demonstrating authorship of the preexisting name should refer to the original publication of the name (but see Note 19.1.1), spelled the same way as when converted and regardless of the rank and composition originally associated with the name (provided it is not a homonym; see Note 9.15A.2). If the original publication of the name cannot be determined, the earliest publication that can be found in which the name is validly published (*ICNAFP*, *ICNP*) or available (*ICZN*) may be cited. If the publication cited is likely not to be the one in which the name was originally published, it should be explicitly stated that the author cited is likely not to be the nominal author (see Art. 19.1) of the name. Under certain conditions (see Notes 9.15A.3 and 9.15A.4), a differently spelled name may be cited. If a citation is for a different spelling than the one adopted in the converted name, the difference in the spelling of the name should be explicitly stated.

Note 9.15A.1. The "original publication of the name" for the purpose of attributing authorship may predate its first publication as a validly published (*ICNAFP*, *ICNP*) or available (*ICZN*) name.

Example 1. Lindley (1830) should be cited as the author of *Angiospermae* under this code even though Lindley's publication of that name was not validly published according to the *ICNAFP* because it was assigned a rank that was contrary to the required relative order of ranks under that code. *Angiospermae* was later validated under the *ICNAFP* by Eichler (1880) and therefore qualifies as a preexisting name under this code (Art. 6.2), but the name is to be attributed under this code to Lindley rather than Eichler.

Note 9.15A.2. In order for two uses of identically spelled preexisting names to be considered the same name rather than homonyms, one use must have been derived from the other or both derived from a third use of the name. If later uses of a name are not accompanied by a reference to an earlier use, absence of any overlap in the compositions associated with identically spelled names can be taken as evidence that they are homonyms (Ex. 1). However, even if there is some overlap, evidence in the protologues may still indicate that the names are homonyms (Ex. 2).

Example 1. If the name *Pholidota* is to be established for a clade of mammals including the pangolins, Weber (1904) should be cited as the author of this name, even though an identically spelled name was published earlier by Merrem (1820). Merrem's (1820) *Pholidota* is considered a homonym, as it was used to refer to a non-overlapping group of organisms later known as *Reptilia*.

Example 2. If the name *Angiospermae* is to be established for the clade comprising the crown clade of flowering plants (or for the clade comprising all flowering plants), Lindley (1830) should be cited as the author of this name, even though an identically spelled name was published earlier by Crantz (1769). Crantz's (1769) *Angiospermae* is considered a homonym even though it was used to refer to a subset of the taxon that Lindley named *Angiospermae*. Crantz's *Angiospermae* was restricted to 13 genera of flowering plants within the clade that is now known as *Lamiales*. Lindley did not refer to Crantz's use of the name, and it is clear that Crantz did not intend the name to refer to all flowering plants.

Note 9.15A.3. For cases in which a preexisting name is attributed to the author of a differently spelled name in the same rank group (e.g., the family group) following the Principle of Coordination of the *ICZN*, that author is not considered under this code to be the author of the preexisting name, nor should the publication of the differently spelled name be cited as an example of use of the preexisting name. The author of the preexisting name is the author of the name as spelled for the purpose of conversion, even if an earlier author who spelled the name differently is considered to be the author of the name under the Principle of Coordination of the *ICZN*, and "the earliest publication that can be found in which the name is validly published (*ICNAFP*, *ICNP*) or available (*ICZN*)" (in Rec. 9.15A) refers only to the converted spelling. However, in such cases, if the earliest author to spell the name as converted is difficult to determine, the person who is considered to be the author of the name under the Principle of Coordination of the *ICZN* may be cited instead, provided that the difference in the spelling of the name is explicitly stated.

Example 1. Under the *ICZN* (1999: Art. 36), Bell is considered to be the author of the name *Iguaninae* because this name was automatically established through the Principle of Coordination when Bell (1825) published *Iguanidae*, even though the first published use of the name *Iguaninae* was by Cope (1886). In contrast, under this code, Cope is considered to be the author of *Iguaninae*. However, if the first author(s) to use the name *Iguaninae* could not be determined, the author could be cited as Bell (1825; as *Iguanidae*).

Note 9.15A.4. For cases in which a preexisting name is attributed to the author of a differently spelled name whose ending has been "corrected" under a rank-based code to the standard ending designated for the rank at which it was published, that author is not considered under this code to be the author of the preexisting name, nor should the publication of the differently spelled name be cited as an example of use of the preexisting name. The author of the preexisting name is the author of the name as spelled for the purpose of conversion, even if an earlier author who spelled the name differently is considered to be the author of the name under the applicable rank-based code, and "the earliest publication that can be found in which the name is validly published (*ICNAFP*, *ICNP*) or available (*ICZN*)" (in Rec. 9.15A) refers only to the converted spelling. However, in such cases, if the earliest author to spell the name as converted is difficult to determine, the person who is considered to be the author of the name under the applicable rank-based code may be cited instead, provided that the difference in the spelling of the name is explicitly stated.

Example 1. Under the *ICNAFP* (Art. 16.3), Jussieu (1789) is considered to be the author of the name *Hypericaceae*, even though he spelled the name *Hyperica*. Under the *ICNAFP*, the name is to be attributed to Jussieu but its spelling is "corrected" to *Hypericaceae*. In contrast, under this code, the author of the name is not considered to be Jussieu, but rather Horaninow (1834) [see Hoogland and Reveal in Bot. Rev. 71: 114 (2005)], who was the first person to publish it with the spelling *Hypericaceae* and in a form that satisfies the other requirements of the *ICNAFP* (see Art. 6.2 of this code). However, if the first author to spell the name *Hypericaceae* could not be determined, the authorship could be cited as Jussieu (1789; as *Hyperica*).

9.16. In order for a bibliographic citation to be direct and unambiguous, it must include author(s)' (see Art. 19) surname(s) and initials, year, title, journal name (where applicable), editors (where applicable), title of the edited book (where applicable), page(s), and plate or figure reference (where applicable). The author(s)' and (where applicable) editor(s)' surname(s) must be cited in full, not abbreviated.

Note 9.16.1. If the protologue or subsequent use of the name to which a bibliographic citation refers is part of a long publication, it may be beneficial to cite (in the text) the page on which the protologue or subsequent use appears, in addition to citing the range of pages of the entire publication (in the bibliography).

Note 9.16.2. If an author or editor does not use a surname, as is true in some cultures, the person's full name as used in the publication should be cited.

Recommendation 9A. Establishment of clade names should be done with careful consideration of possible nomenclatural consequences if the phylogenetic hypothesis turns out to be incorrect. It may frequently be advisable to use only informal names for poorly supported clades.

Recommendation 9B. Conversion of preexisting names to clade names should only be done with a thorough knowledge of the group concerned, including its taxonomic and nomenclatural history and previously used diagnostic features. Wholesale conversion of preexisting names by authors who have not worked on the systematics of the group concerned is strongly discouraged.

Recommendation 9C. In order to facilitate the referral of species and specimens that are not specifiers of the clade name, as well as less inclusive clades, the protologue should include a description, diagnosis, or list of apomorphies.

Note 9C.1. A diagnosis or description is required for simultaneous valid publication (*ICNAFP*, *ICNP*) or availability (*ICZN*) of the name under the appropriate rank-based code.

ARTICLE 10. SELECTION OF CLADE NAMES FOR ESTABLISHMENT

10.1. Clade names are generally to be selected in such a way as to minimize disruption of current and/or historical usage (with regard to composition, diagnostic characters, or both) and to maximize continuity with existing literature. Therefore, except under the conditions described in Article 10.2, a preexisting name that has been applied to a taxon approximating the clade to be named (see Note 9.15.2), or to a paraphyletic group originating in the same ancestor, must be selected. If there is a preexisting name for a paraphyletic group originating in the same ancestor as a particular clade and that name is much better known than any preexisting name for that clade, or if there is no preexisting name for that clade, the name of the paraphyletic group may be (but need not be) chosen.

Note 10.1.1. Article 10.1 and Recommendation 10.1A are not intended either to encourage or to discourage the application of preexisting names to crown, apomorphy-based, or total clades. Because the associations of preexisting names with precisely identified clades commonly are ambiguous, reasonable arguments can often be made for applying a particular name to any one of several nested clades between crown and total (inclusive).

Recommendation 10.1A. If more than one preexisting name has been applied to a particular clade (including those applied to paraphyletic groups originating in the same ancestor), the name that is most widely and consistently used for it should generally be chosen, though a less widely used name may be chosen if it is a panclade name (see Art. 10.3 and Note 10.3.1). Similarly, if a preexisting name has been applied to more than one clade, it should generally be established for the clade to which it has been most widely and consistently applied (but see Note 10.1.1). If the most widely and consistently used name is not selected for conversion, a rationale should be provided.

40

Note 10.1A.1. In selecting "the name that is most widely and consistently used," considerable discretion is left to the converting author. It is not necessary to choose a name that is slightly more widely used than its closest competitor. As a general guideline, if there is less than a twofold difference in the frequency of use of two or more names, the converting author may choose any of them without providing a compelling justification.

Recommendation 10.1B. The name that is more commonly used than any other name to refer to (e.g., discuss or describe) a particular crown clade should generally be defined as applying to that crown clade, even if the name is commonly considered to apply to a clade that includes extinct taxa outside of the crown. If there is a conflict between Recommendations 10.1A and 10.1B, Recommendation 10.1B should be given precedence. If the name that is more commonly used than any other name to refer to a crown clade is instead defined as applying to a larger clade (e.g., an apomorphy-based or total clade) that contains that crown, a justification should be provided.

Note 10.1B.1. In older works and in works dealing only with extant organisms, names have sometimes been used as if they apply to particular crown clades, though it is unclear whether the author considered the name to apply to the crown or to a larger clade (i.e., including some or all of the stem). In such cases, the name may be interpreted as applying to the crown for the purpose of this recommendation.

Example 1. If a publication stated that all members of clade X (e.g., *Mammalia*) exhibit a particular feature M (e.g., lactation), and this feature has only been observed in extant species, the name X would have been used in that publication as if it applied to the crown clade. Given this situation, name X could be interpreted as a candidate name for the crown.

10.2. A new name may be selected for a clade only under one of the following circumstances: (a) the clade has no preexisting name (but see Note 10.2.1); (b) the most widely used preexisting name for the clade has already been established for another clade or is best applied to another clade (see Recs. 10.1A and 10.1B), and there are no other preexisting names for the clade; (c) the most widely used preexisting name for the clade has a preexisting homonym that has already been established under this code (see Recs. 10D–F); (d) the group to be named is a total clade, in which case a panclade name (see Arts. 10.3–10.7) may be used instead of a preexisting name; (e) the name is to be given an apomorphy-based definition, and the name of the largest crown clade that inherited the apomorphy of concern refers etymologically to that apomorphy (see Arts. 10.8–10.9).

Note 10.2.1. In the absence of a preexisting name for a particular clade, the choice between a new name and a preexisting name for a paraphyletic group originating in the same ancestor as the clade is left to the discretion of the author.

10.3. If a new name (as opposed to a converted name) is to be established for a total clade by adding an affix to the name of the corresponding crown clade, the prefix *Pan-* must be used. The prefix is separated from the base name, which retains an initial capital letter, by a hyphen. Such names are called panclade names and may only be used to designate total clades.

Example 1. If *Testudines* is established as the name of a crown clade, the panclade name for the corresponding total clade is *Pan-Testudines*.

Note 10.3.1. Although most panclade names will be new, some panclade names may have been defined in a publication prior to the starting date of this code, and these qualify as preexisting names (see Note 6.2.1).

Recommendation 10.3A. Informal panclade names may be useful in referring to total clades that may or may not have formal (established) panclade names. In order to distinguish informal panclade names from formal panclade names, the informal names should not be capitalized or italicized (see Rec. 6.1A).

Example 1. The non-capitalized, non-italicized names pan-Rosidae and pan-rosids are informal panclade names for the total clade corresponding to the crown clade *Rosidae*.

10.4. A panclade name may only be formed from a base name that has a crown-clade definition (see Art. 9.9).

Example 1. If the names *Trilobita* and *Tyrannosaurus* were established as the names of non-crown clades, then the names *Pan-Trilobita* and *Pan-Tyrannosaurus* could not be established as clade names.

Recommendation 10.4A. Some converted clade names will necessarily begin with *Pan*, but the initial letters *Pan* should be avoided in new clade names that are not intended as panclade names to reduce the likelihood of confusion between panclade and non-panclade names.

10.5. The definition of a panclade name is formulated through reference to its corresponding crown clade and will take the form "the total clade composed of the crown clade [name of the crown clade] and all extinct organisms or species that share a more recent common ancestor with [name of the crown clade] than with any extant organisms or species that are not members of [name of crown clade]" or "the total clade of the crown clade [name of the crown clade]". The definition of a panclade name may be abbreviated "total V of X", where X is the name of a crown clade.

Example 1. The definition of *Pan-Testudines* is "the total clade composed of the crown clade *Testudines* and all extinct organisms or species that share a more recent common ancestor with *Testudines* than with any extant organisms or species that are not members of *Testudines*" or "the total clade of the crown clade *Testudines*". The abbreviated form of this definition is "total ∇ of *Testudines*".

Note 10.5.1. The definition of a panclade name may alternatively be abbreviated "total ∇ (X)", where X is the name of a crown clade. In this case, () = of (rather than "containing" or "contain" as in other abbreviated definitions; see Notes 9.4.1 and 11.12.1).

Example 1. The alternative abbreviated form of the definition in Article 10.5, Example 1 is "total ∇ (*Testudines*)".

Note 10.5.2. This format for the definitions of panclade names differs from the other recommended definition formats (see Arts. 9.5–9.7 and 9.9–9.10) in not listing any specifiers, which are implicit. The internal specifiers of the panclade name are those of the crown-clade name on which the panclade name is based. The external specifiers of the panclade name are all extant species or organisms that are not members of the crown clade on whose name the panclade name is based.

Note 10.5.3. Extinction of crown clades after establishment of a panclade name does not affect the composition of the clade to which the panclade name refers. A crown clade that is extant at the time of establishment of a panclade name is forever treated nomenclaturally as though it were still extant. This treatment applies both to the crown clade that provides the base name for the panclade name and to other crown clades whose members are implicit external specifiers (see Note 10.5.2).

10.6. If there is a preexisting name that has been applied to a particular total clade, that name may be converted or a panclade name may be established instead.

10.7. A panclade name can only be the accepted name of the total clade to which it applies if the crown-clade name upon which it is based is the accepted name of the corresponding crown clade, even if no other name has been established for that total clade (see also Art. 14.5).

Example 1. Suppose that: (1) the name *Pan-Lepidosauria* has been established for the total clade of the crown clade for which the name *Lepidosauria* has been established, and (2) a panclade name has not been established for the total clade of the crown clade for which the name *Reptilia* has been established, and (3) *Reptilia* has precedence over *Lepidosauria*. In the context of phylogenies in which the names *Reptilia* and *Lepidosauria* are synonyms, *Reptilia* would be the accepted name of the crown clade. Although the name *Pan-Lepidosauria* has been established and its definition indicates application to the corresponding total clade, it cannot be the accepted name of that total clade because the name on which it is based, *Lepidosauria*, is not the accepted name of the corresponding crown clade.

10.8. If the name of a crown clade refers etymologically to an apomorphy, and a new name (as opposed to a converted name) is to be established for the clade originating with that apomorphy by adding an affix to the name of the crown clade, the prefix *Apo-* must be used. The prefix is separated from the base name, which retains an initial capital letter, by a hyphen.

Example 1. In *Phylonyms*, Cantino et al. defined *Spermatophyta* (meaning "seed plants") as the name of a crown clade, and they defined *Apo-Spermatophyta* to apply to the larger clade for which "seeds" is an apomorphy.

Note 10.8.1. Although most names that take the form described in Article 10.8 will be new, some such names may have been defined in a publication prior to the starting date of this code, and these qualify as preexisting names (see Note 6.2.1).

10.9. If there is a preexisting name that has traditionally been applied to a clade characterized by a particular apomorphy, and if a different name that refers etymologically to that apomorphy has been established for the largest crown clade that inherited the apomorphy of concern, the preexisting name may be converted with an apomorphy-based definition for the former clade or a name formed in accordance with Article 10.8 may be established instead.

10.10. A clade name may not be converted from a preexisting specific or infraspecific epithet (*ICNAFP* and *ICNP*) or from a name in the species group (*ICZN*). However, a clade name may be converted from a supraspecific name that is spelled identically to a specific or infraspecific epithet or name.

Example 1. A clade cannot take the name *Paradoxa* if the name was converted from the specific epithet in *Oenothera paradoxa* Hudziok 1968; however, a clade can take the name *Paradoxa* if the name was converted from the genus name *Paradoxa* Mattirolo 1935.

Recommendation 10A. In selecting new clade names, an effort should be made to avoid any name that, under a rank-based code, applies to a non-overlapping (mutually exclusive) group.

Recommendation 10B. In selecting new clade names, an effort should be made to avoid names that are so similar to names that were previously established under this code that they are likely to be confused.

Recommendation 10C. In selecting new clade names, an effort should be made to avoid names that have misleading connotations.

Recommendation 10D. In rank-based nomenclature, there are many examples of identically spelled names being applied to different taxa under different codes (cross-code homonyms). Only one member of each set of cross-code homonyms is, after conversion, an acceptable name under this code (Art. 13.3). If the preexisting name that has been most widely used for a particular clade cannot be converted because an identically spelled name has already been converted and established for a different clade, another preexisting name that has been widely and recently applied to the clade concerned (or to a paraphyletic group originating in the same ancestor) may be selected. On the other hand, continuity with existing literature and consistency with rank-based nomenclature are not well served by resurrecting old and little-known names. Therefore, if there is no other name that has been widely applied to the clade in the recent past, a new name should be selected that consists of the most widely used preexisting name with a taxon-related prefix added, such as *Phyto-* for plants, *Phyco-* for "algae" excluding cyanobacteria, and *Myco-* for fungi (three groups of organisms whose names are governed by the *ICNAFP*), *Zoo-* for animals, *Protisto-* for other organisms whose names are governed by the *ICZN* (i.e., non-photosynthetic "protists"), and *Prokaryo-* or *Bacterio-* for organisms whose names are governed by the *ICNP* and cyanobacteria (governed by the *ICNAFP*). The prefix is separated from the preexisting name, which retains an initial capital letter, by a hyphen. If there is another preexisting name that has been widely applied to the clade in the recent past, the choice between converting this name and establishing a new name with a taxon-related prefix is left to the discretion of the author.

Example 1. Under rank-based nomenclature, the name *Prunella* applies to a genus of birds (*ICZN*) and to a genus of angiosperms (*ICNAFP*). If this name were to be established under this code for a clade of birds, the name selected for the clade corresponding in composition to the plant genus *Prunella* (provided that there is no other preexisting name that has been widely and recently applied to this clade) would be *Phyto-Prunella*.

Recommendation 10E. In rank-based nomenclature, previously undiscovered homonymy occasionally occurs within a single rank-based code, although only one of the homonyms can be legitimate (*ICNAFP, ICNP*) or potentially valid (*ICZN*) once the homonymy is discovered. Only one member of each set of homonyms is, after conversion, an acceptable name under this code (Art. 13.3). Once a case of homonymy within a rank-based code is discovered, it is generally rectified by replacing the later homonym with an already existing synonym or a new replacement name. However, if a user of this code is the first to discover a case of homonymy within one of the rank-based codes, the names should be defined in a manner that is consistent with the way in which they will likely be applied under the rank-based code when the situation is rectified. Specifically, the homonym that will likely have precedence under the rank-based code (i.e., generally the one that was published earlier) should be the one that is converted under this code. For the other homonym, the synonym (if one exists) that will likely be applied to this taxon under the rank-based code should be converted, provided that this synonym qualifies as a preexisting name for the clade of concern (see Art. 9.15).

Note 10E.1. In the situation described in Recommendation 10E, it is not necessary that an author who converts one homonym (or its synonym) also convert the other one (or its synonym).

Recommendation 10F. Under rank-based nomenclature, the name (or epithet; see below) of a subdivision of a genus that contains the type species must be the same as that of the genus. Only one member of each such pair of names is, after conversion, an acceptable name under this code (Art. 13.3). Furthermore, under the *ICNAFP*, names of subdivisions of genera (e.g., subgenera, sections, series) consist of a generic name combined with a subdivisional epithet. These epithets, like specific epithets, are not necessarily unique; the same epithet may be combined with the names of different genera without creating homonyms. Only one member of each set of identically spelled subdivisional epithets is, after conversion, an acceptable name under this code (Art. 13.3). If the preexisting subdivisional name (*ICZN, ICNP*) or epithet (*ICNAFP*) that has been most widely used for a particular clade cannot be converted because an identically spelled name has already been converted and established for a different clade, another preexisting name or epithet that has been widely and recently applied to the clade concerned (or to a paraphyletic group originating in the same ancestor) may be selected. On the other hand, continuity with existing literature and consistency with rank-based nomenclature are not well served by resurrecting old and little-known names. Therefore, if there is no other name or epithet that has been widely applied to the clade in the recent past, a new name should be selected that consists of the most widely used preexisting name or epithet, preceded by the name of the genus in rank-based nomenclature, with both words capitalized and connected by a hyphen. If there is another preexisting name or epithet that has been widely applied to the clade in the recent past, the choice between converting this name or epithet and establishing a new name that combines the preexisting genus name and subdivisional name or epithet is left to the discretion of the author.

Example 1. If one were selecting a name for the plant clade corresponding in composition to *Arenaria* sect. *Parviflorae* McNeill, and if the subdivisional epithet *Parviflorae* could not be converted because a clade name *Parviflorae*, based on *Dracula* ser. *Parviflorae* Luer, had already been established under this code, the name that should be selected is *Arenaria-Parviflorae* (provided that there is no other preexisting name that has been widely and recently applied to this clade).

Note 10F1.1. This is a hypothetical example in that these subdivisions of genera may not correspond to clades.

Example 2. If one were selecting a name for the animal clade corresponding in composition to the subgenus *Crotaphytus* of the genus *Crotaphytus* Holbrook, and if the name *Crotaphytus* could not be converted for that clade because that name had already been established under this code for a clade corresponding in composition with the genus, the name that should be selected is *Crotaphytus-Crotaphytus* (provided that there is no other preexisting name that has been widely and recently applied to this clade).

Recommendation 10G. When establishing a name for a crown clade that, under rank-based nomenclature, corresponds to a monogeneric "higher" taxon, the genus name should be converted for that clade rather than any of the suprageneric names that have been applied to it. Doing so will permit the use of the "higher" taxon names for larger clades that extend beyond the crown.

Example 1. In rank-based nomenclature, the names *Equisetophyta*, *Equisetopsida*, *Equisetales*, *Equisetaceae*, and *Equisetum* have all been used to refer to the same crown clade, which is widely understood to include only the genus *Equisetum*. (Most of these names have also been used to refer to more inclusive clades that contain extinct species outside the crown.) When selecting a name to convert for the crown clade, *Equisetum* should be chosen. The names *Equisetaceae*, *Equisetales*, etc. are better applied to clades that are more inclusive than the crown.

ARTICLE 11. SPECIFIERS AND QUALIFYING CLAUSES

11.1. Specifiers are species, specimens, or apomorphies cited in a phylogenetic definition of a name as reference points that serve to specify the clade to which the name applies. All specifiers used in minimum-clade and maximum-clade definitions, and one of the specifiers used in apomorphy-based definitions, are species or specimens. The other specifier used in an apomorphy-based definition is an apomorphy.

Note 11.1.1. Although subordinate clades cannot be specifiers, they may be cited in a phylogenetic definition of the name of a more inclusive clade to clarify the phylogenetic position of a specifier.

Example 1. *Aves* could be defined as "the crown clade originating in the most recent common ancestor of *Struthio camelus* Linnaeus 1758 (*Palaeognathae*) and *Vultur gryphus* Linnaeus 1758 (*Neognathae*)." Alternatively, the definition could be worded "the crown clade originating in the most recent common ancestor of *Palaeognathae* (*Struthio camelus* Linnaeus 1758) and *Neognathae* (*Vultur gryphus* Linnaeus 1758)." In both definitions, *Palaeognathae* and *Neognathae* are not specifiers; they simply provide additional information about the phylogenetic position of the true specifiers.

11.2. An internal specifier is a species, specimen, or apomorphy that is explicitly included in the clade whose name is being defined; an external specifier is a species or specimen that is explicitly excluded from it. All specifiers in minimum-clade (including minimum-crown-clade), apomorphy-based, and apomorphy-modified crown-clade definitions are normally internal (though external specifiers may be used to prevent use of a name under certain hypotheses of relationships, clade composition, or both; see Art. 11.13, Ex. 1), but maximum-clade (including maximum-crown-clade and maximum-total-clade) definitions always have at least one specifier of each type.

11.3. When a species is used as a specifier, the author and publication year of the species name must be cited.

Note 11.3.1. Names of species used as specifiers are governed by the rank-based codes (e.g., *ICNAFP*, *ICZN*); see Article 21.

Note 11.3.2. The *ICNAFP* and *ICZN* differ in their conventions for citing authorship and publication year when the author of the currently accepted binominal combination differs from the author of the epithet (i.e., the author of the original combination). Because the purpose of citing authorship and year is to identify the specifiers unambiguously, the conventions used by the appropriate rank-based code should be used for species names governed by that code.

11.4. Because species names are governed by the rank-based codes, they are generally associated with type specimens, which serve as reference points for the names; this situation bears on the use of species as specifiers under this code. In effect, whichever currently accepted species includes the type specimen of the species name cited in the definition is the specifier. If the species name originally cited in a definition is no longer accepted, then the species name with which it has been synonymized becomes the name of the specifier species. However, should the species name originally cited in the definition become the accepted name of a species at a later time, that name would again become the name of the specifier species. The status of a species as a specifier is unaffected by the designation of a new or different type under the appropriate rank-based code. A species may be used as a specifier even if its name lacks a type. See also Recommendation 11H, regarding the naming of low-level clades.

11.5. When a type specimen is used as a specifier, the species name that it typifies and the author and publication year of that species name must be cited.

Recommendation 11.5A. When type specimens are used as specifiers in the definitions of the names of low-level clades (see Rec. 11H), holotypes or lectotypes should be used. If there is no holotype or lectotype that, if designated as a specifier, would result in the name being applied to the intended clade, and a syntype is selected instead, it is recommended that the same syntype be simultaneously designated as the lectotype under the appropriate rank-based code.

11.6. When a type specimen is used as a specifier, it retains its status as a specifier even if a different type for the species name that it typified is subsequently designated under the relevant rank-based code, or if the species name that it typifies is no longer accepted because the species has been re-circumscribed and the name relegated to synonymy.

11.7. Specimens that are not types may not be used as specifiers unless: (1) the specimen that one would like to use as a specifier cannot be referred to a named species, so that there is no type specimen that could be used instead; or (2) the clade to be named is nested entirely within a species; or (3) the clade to be named includes part of a non-monophyletic species and its descendants, but the type of the non-monophyletic species is either excluded from that clade or it is not possible to determine whether it is included.

Note 11.7.1. There may be differences in taxonomic opinion as to whether a specimen can be referred to a named species (see Art. 11.7: situation 1). Article 11.7 is not intended to provide a basis for challenging the establishment of a clade name by adopting a different species definition from that of the definitional author and thereby challenging the claim that a particular specifier cannot be referred to a named species. However, wholesale rejection of the species category cannot be used as a basis for claiming that a particular specimen cannot be referred to any species, and the use of non-types as specifiers based on such an argument constitutes grounds for challenging the establishment of the name.

Recommendation 11.7A. If a specimen that is not a type is used as a specifier in the first situation described in Article 11.7, and a species that includes this specimen is subsequently named under the appropriate rank-based code, this specimen should be chosen as the type of the species name.

11.8. When a specimen that is not a type is used as a specifier in a phylogenetic definition, the institution or collection in which the specifier is conserved must be identified, as well as the collection number or other information needed to identity the specimen unambiguously.

11.9. When a specimen that is not a type is used as a specifier in a phylogenetic definition, either a brief description or an image or reference to a published image of the specimen must be provided, sufficient to convey a mental image to a non-specialist and distinguish the specimen from organisms with which it might be confused. If a copyrighted image is submitted, the registration database (see Art. 8.1) must be provided with written authorization to share it with database users. If a reference to a published image is submitted, it must include all the information necessary to locate it.

11.10. In the interest of avoiding confusion, a clade name should not be based on the name of another taxon that is not part of the named clade. Therefore, when a clade name is converted from a preexisting name that is typified under a rank-based code or is a new or converted name derived from the stem of a typified name, the definition of the clade name must use the type species of that preexisting typified name or of the genus name from which it is derived (or the type specimen of that species) as an internal specifier.

Example 1. If the preexisting name *Magnoliales*, which is based on the genus name *Magnolia*, is converted to a clade name, its definition must use the type species of *Magnolia* or its type specimen as an internal specifier.

Example 2. If *Ajugina*, which is not a preexisting name but is based on the preexisting genus name *Ajuga*, is adopted as the name of a clade, the definition of *Ajugina* must use the type species of *Ajuga* or its type specimen as an internal specifier.

Example 3. If the preexisting subgenus name *Calosphace* is converted to a clade name, the definition of *Calosphace* must use the type species of subgenus *Calosphace* or its type specimen as an internal specifier.

Example 4. If the preexisting name *Caprifoliaceae*, which is based on the genus name *Caprifolium*, is converted to a clade name, its definition must use the type species of *Caprifolium* or its type specimen as an internal specifier. This is true even though the name *Caprifolium* is a later synonym of *Lonicera* under the *ICNAFP* and therefore is not an accepted genus name.

Recommendation 11.10A. If it is questionable whether a type species of a preexisting name is part of the clade to be named, then the type species should not be used as a specifier (see Rec. 11B), and neither that preexisting name nor a name derived from the stem of that name should be defined as referring to that clade.

Example 1. If it is questionable whether the type species of *Magnolia* belongs to a clade that is to be named, this species should not be used as a specifier, and the clade should not be named *Magnolia*, *Magnoliales*, or any other name based on the stem of the name *Magnolia*.

Note 11.10A.1. Failure to include the type species of a preexisting name in an analysis is not, in itself, reason to invoke Recommendation 11.10A. There may be evidence suggesting that another species that was included in the analysis shares a recent common ancestor with the type.

Recommendation 11.10B. If it is questionable whether the type specimen of a preexisting name belongs to the clade to be named (e.g., because of the fragmentary nature of the specimen), then neither that specimen nor the species name that it typifies should be used as a specifier (see Rec. 11C), and the corresponding name should not be converted to a clade name.

Example 1. Under the *ICNAFP*, the names *Cordaites*, *Cordaixylon*, and *Mesoxylon* refer to genera of extinct seed plants. The types of the latter two names are fossil stems, but it has been possible to reconstruct whole plants that belonged to each genus. The oldest of the three names, *Cordaites*, is typified by fossil leaf material that could have been produced by a member of either *Cordaixylon* or *Mesoxylon*. If a clade is named that includes plants with *Cordaixylon*-type stems but not *Mesoxylon*-type stems, neither the type specimen of *Cordaites* nor the species name that it typifies should be cited as a specifier because they may not belong to this clade, and the clade should not be named *Cordaites*.

11.11. In order for the reference phylogeny to be useful, either the specifiers used in the phylogenetic definition must be included in the reference phylogeny (see Art. 9.13) or the protologue must include a statement indicating how the specifiers are related to the taxa that are included in the reference phylogeny.

Note 11.11.1. An acceptable mechanism for indicating how the specifiers are related to the taxa in the reference phylogeny is to cite a taxon name that is labeled on the reference phylogeny parenthetically after the name of the specifier in the definition.

Note 11.11.2. If a clade name is converted from a typified name under a rank-based code, or is derived from the stem of such a name, and the type of that name is not included in the reference phylogeny, the type must still be used as an internal specifier (see Art. 11.10), and its relationship to taxa included in the reference phylogeny must be stated in the protologue.

11.12. In order to prevent use of a name under certain hypotheses of relationships, clade composition, or both, phylogenetic definitions may include qualifying clauses specifying conditions under which the name cannot be applied to any clade (see Exs. 1 and 2).

Note 11.12.1. The following conventions are adopted for abbreviated qualifying clauses such as those in Examples 1 and 2: | = on the condition that; ˜ = it does not; () = contain; ∨ = or; and anc = the ancestor in which the clade originated. See Note 9.4.1 for the other abbreviations used in these examples.

Example 1. The name *Pinnipedia* is traditionally applied to a group composed of sea lions (*Otariidae*), walruses (*Odobenidae*), and seals (*Phocidae*). However, under some phylogenetic hypotheses, the sister group of one or more of these taxa is a group of terrestrial carnivorans (e.g., *Ursidae, Procyonidae, Mustelidae*). If the name *Pinnipedia* were to be defined as "the clade originating in the most recent common ancestor of *Otaria byronia* de Blainville 1820, *Odobenus rosmarus* Linnaeus 1758, and *Phoca vitulina* Linnaeus 1758, provided that it does not include *Ursus arctos* Linnaeus 1758 or *Procyon lotor* (Linnaeus 1758) or *Mustela erminea* Linnaeus 1758", then the name would not be applicable to any clade in the context of phylogenetic hypotheses in which the most recent common ancestor of *Otaria byronia, Odobenus rosmarus,* and *Phoca vitulina* was also inferred to be an ancestor of *Ursus arctos* or *Procyon lotor* or *Mustela erminea*. The phrase "provided that it does not include *Ursus arctos* Linnaeus 1758 or *Procyon lotor* (Linnaeus 1758) or *Mustela erminea* Linnaeus 1758" is a qualifying clause. This definition may be abbreviated min ∨ (*Otaria byronia* de Blainville 1820 & *Odobenus rosmarus* Linnaeus 1758 & *Phoca vitulina* Linnaeus 1758) | ˜ (*Ursus arctos* Linnaeus 1758 ∨ *Procyon lotor* (Linnaeus 1758) ∨ *Mustela erminea* Linnaeus 1758) (see Note 11.12.1).

Example 2. The name *Pinnipedia* is traditionally applied to a group composed of sea lions (*Otariidae*), walruses (*Odobenidae*), and seals

(*Phocidae*). However, under some phylogenetic hypotheses, the sister group of one or more of these taxa is a group of terrestrial carnivorans. If the name *Pinnipedia* were to be defined as "the clade originating in the most recent common ancestor of *Otaria byronia* de Blainville 1820, *Odobenus rosmarus* Linnaeus 1758, and *Phoca vitulina* Linnaeus 1758, provided that the ancestor in which the clade originated possessed flippers synapomorphic with those in the aforementioned species," then the name would not be applicable to any clade in the context of phylogenetic hypotheses in which the most recent common ancestor of these species was inferred not to have had flippers. The phrase "provided that it possessed flippers synapomorphic with those in the aforementioned species" is a qualifying clause. (However, the apomorphy "flippers" should be illustrated or described because it is a complex apomorphy (see Recs. 9.7A, 9.7B).) This definition may be abbreviated min ∇ (*Otaria byronia* de Blainville 1820 & *Odobenus rosmarus* Linnaeus 1758 & *Phoca vitulina* Linnaeus 1758) | anc possessed flippers synapomorphic with those in the aforementioned species (see Note 11.12.1).

11.13. The use of a name under certain hypotheses of relationships, clade composition, or both can also be prevented by using a minimum-clade definition with external specifiers (Ex. 1) or a maximum-clade definition with more than one internal specifier (Ex. 2) or an apomorphy-based definition with more than one internal specifier (Ex. 3). These definitions have the same effect as qualifying clauses (Art. 11.12) in that under some phylogenetic hypotheses, the name cannot be applied to any clade.

Example 1. If a name is defined through a minimum-clade definition (or a minimum-crown-clade definition) with an external specifier, and one internal specifier is later found to share a more recent common ancestor with the external specifier than with the other internal specifier, the definition does not apply to any clade. For example, suppose the name *Halecostomi* had been defined as referring to the smallest clade containing *Amia calva* Linnaeus 1766 and *Perca fluviatilis* Linnaeus 1758 but not *Lepisosteus osseus* Linnaeus 1758. And suppose that subsequent analyses indicated that *Lepisosteus osseus* and *Perca fluviatilis* share a more recent common ancestor with one another than either does with *Amia calva*. If so, then there would be no clade that fits the definition of *Halecostomi* (because there would be no clade that includes both *Amia calva* and *Perca fluviatilis* but not *Lepisosteus osseus*), and that name could not be used in the context of that hypothesis.

Example 2. If a name is defined through a maximum-clade definition (or a maximum-crown-clade definition or a maximum-total-clade definition) with more than one internal specifier, and one internal specifier is later found to share a more recent common ancestor with the external specifier than with the other internal specifier, the definition does not apply to any clade. For example, suppose the name *Halecostomi* had been defined as referring to the largest clade containing *Amia calva* Linnaeus 1766 and *Perca fluviatilis* Linnaeus 1758 but not *Lepisosteus osseus* Linnaeus 1758. And suppose that subsequent analyses indicated that *Lepisosteus osseus* and *Perca fluviatilis* share a more recent common ancestor with each other than either does with *Amia calva*. If so, then there would be no clade that fits the definition of *Halecostomi* (because there would be no clade that includes both *Amia calva* and *Perca fluviatilis* but not *Lepisosteus osseus*), and that name could not be used in the context of that hypothesis.

Example 3. If a name is defined through an apomorphy-based definition with multiple internal specifiers, and it is later found that the apomorphy is not homologous in all of the internal specifiers, the definition does not apply to any clade. For example, Cantino, Doyle, and Donoghue defined *Apo-Spermatophyta* in *Phylonyms* as the clade characterized by seeds as inherited by *Magnolia tripetala* (Linnaeus) Linnaeus 1759, *Podocarpus macrophyllus* (Thunberg) Sweet 1818, *Ginkgo biloba* Linnaeus 1771, *Cycas revoluta* Thunberg 1782, and

Gnetum gnemon Linnaeus 1767. Suppose that subsequent analyses indicated that the seeds of cycads arose separately from those of the other specifiers (though this hypothesis is not supported by any modern analysis). If so, then there would be no clade that fits the definition, and the name *Apo-Spermatophyta* could not be used in the context of that hypothesis.

11.14. The application of a name with respect to clade composition under alternative hypotheses of relationship can be restricted by defining it relative to the name of another clade (Ex. 1). However, unlike the mechanisms described in Articles 11.12 and 11.13, the name does not become inapplicable under the alternative phylogenetic hypothesis.

Example 1. Gauthier et al. (1988) proposed the name *Lepidosauriformes* (max ∇ (*Lacerta agilis* Linnaeus 1758 ~ *Youngina capensis* Broom 1914) for a subclade of *Lepidosauromorpha* (max ∇ (*Lacerta agilis* Linnaeus 1758 ~ *Crocodylus niloticus* Laurenti 1768)), which was itself proposed for a subclade of *Sauria* (min ∇ (*Lacerta agilis* & *Crocodylus niloticus*)). If *Youngina capensis* turned out to be outside of the clade originating in the most recent common ancestor of *Lacerta agilis* and *Crocodylus niloticus* (i.e., *Sauria*), then the name *Lepidosauriformes* would refer to a larger clade than *Lepidosauromorpha*, reversing the former hierarchical relationships of the clades designated by those names. To prevent that reversal, the name *Lepidosauriformes* could have been defined as "the largest clade within *Sauria* containing *Lacerta agilis* but not *Youngina capensis*," in which case *Lepidosauriformes* would become a synonym of *Lepidosauromorpha* (rather than the name of a larger clade) in the context of the new phylogenetic hypothesis. In contrast to the original definition, the addition of "within *Sauria*" in the alternative definition restricts application of the name to a subclade of *Sauria*. (Note that a similar restriction could be achieved by using *Crocodylus niloticus* as an additional external specifier.)

11.15. Provided that a clade name is acceptable, it remains eligible for use even if there is no clade that fits its definition under a subsequently proposed phylogenetic hypothesis. The name would not be used in the context of that hypothesis, but it would remain eligible for future use under any hypotheses in which there is a clade that fits its definition.

Example 1. Although the name *Pinnipedia* is inapplicable under certain phylogenetic hypotheses if the qualifying clause in Article 11.12, Example 1 is used, the name remains eligible for use under other hypotheses.

Recommendation 11A. Definitions of converted clade names should be stated in a way that attempts to capture the spirit of traditional use to the degree that it is consistent with the contemporary concept of monophyly. Consequently, such a definition should not necessitate (though it may allow) the inclusion of subtaxa that have traditionally been excluded from the taxon designated by the preexisting name, as well as the exclusion of subtaxa that have traditionally been included in the taxon. To accomplish this goal, internal specifiers of converted clade names should be chosen from among the taxa that have been considered to form part of a taxon under traditional ideas about the composition of that taxon, and they should not include members of subtaxa that have traditionally been considered not to be part of the taxon.

Example 1. The name *Dinosauria* was coined by Owen for the taxa *Megalosaurus*, *Iguanodon*, and *Hylaeosaurus*, and traditionally the taxon designated by that name has included these and certain other non-volant reptiles. It has not traditionally included birds. Although birds are now considered part of the dinosaur clade, the name *Dinosauria* should not be defined using any bird species as internal specifiers. Such a definition would force birds to be dinosaurs, thus trivializing the question of whether birds are dinosaurs. Instead, internal specifiers should be chosen from among taxa that have traditionally been considered dinosaurs, e.g., *Megalosaurus bucklandii* Mantell 1827, *Iguanodon bernissartensis* Boulenger in Beneden 1881, and *Hylaeosaurus armatus* Mantell 1833.

Note 11A.1. Traditional use may refer to early or recent traditions. Because it is not always possible to be faithful to all traditions simultaneously, which tradition is most important to maintain is left to the discretion of the author of the converted name.

Recommendation 11B. If there is reason to question that a species is a member of a particular clade, that species should not be used as a specifier in the definition of the name of that clade.

Recommendation 11C. Because they are commonly based on taxonomically ambiguous types, ichnotaxa (taxa based on the fossilized work of organisms, including fossilized trails, tracks, and burrows; *ICZN* (1999) glossary, Art. 1.2.1) and ootaxa (taxa based on fossilized eggs) should not be used as specifiers. When this recommendation is combined with Article 11.10, it follows that clade names should not be based on the names of ichnotaxa or ootaxa.

Recommendation 11D. In a minimum-clade definition, it is best to use a set of internal specifiers that includes representatives of all subclades that credible evidence suggests may be sister to the rest of the clade being named, unless doing so would be contrary to Recommendation 11A and/or 11B. Constructing a minimum-clade definition in this way will reduce the chance that, under a new phylogenetic hypothesis, the name will refer to a less inclusive clade than originally intended.

Recommendation 11E. In a maximum-clade definition, it is best to use a set of external specifiers that includes representatives of all clades that credible evidence suggests may be the sister group of the clade being named. Constructing a maximum-clade definition in this way will reduce the chance that, under a new phylogenetic hypothesis, the name will refer to a more inclusive clade than originally intended.

Recommendation 11F. If it is important to establish two names as applying to sister clades regardless of the phylogeny, reciprocal maximum-clade definitions should be used in which the single internal specifier of one is the single external specifier of the other, and vice versa. To establish a name as applying to the larger clade composed of those two sister-clades, the name of the former should be given a minimum-clade definition using the same two internal specifiers (Ex. 1). A similar approach may be used to establish two names as referring to crown clades that are each other's closest extant relatives by using reciprocal maximum-crown-clade definitions (Ex. 2).

Example 1. If one wishes to define the names *Saurischia* and *Ornithischia* such that they will always refer to sister clades, *Saurischia* might be defined as the largest clade containing *Megalosaurus bucklandii* von Meyer 1832 but not *Iguanodon bernissartensis* Boulenger in Beneden 1881, and *Ornithischia* would be defined as the largest clade containing *Iguanodon bernissartensis* but not *Megalosaurus bucklandii*. To stabilize the name *Dinosauria* as referring to the clade comprising *Saurischia* and *Ornithischia*, *Dinosauria* should be defined as the smallest clade containing *Megalosaurus bucklandii* and *Iguanodon bernissartensis*.

Example 2. If one wishes to define the names *Lamioideae* and *Scutellarioideae* such that they will always refer to crown clades that are each other's closest extant relatives, *Lamioideae* might be defined as the largest crown clade containing *Lamium album* L. 1753 but not *Scutellaria galericulata* L. 1753, and *Scutellarioideae* would be defined as the largest crown clade containing *Scutellaria galericulata* but not *Lamium album*.

Recommendation 11G. Clade names (new or converted) that combine certain prefixes or suffixes with another clade name (the base name) should be defined in a manner consistent with the hierarchical relationships implied by the prefix or suffix and the phylogenetic definition of the base name (if established), unless doing so would be inconsistent with the predominant current use of a preexisting name (see also Note 11G.2).

Example 1. If preexisting names *Parahebe* and *Hebe* are converted, the internal specifiers of each name should not include any member of the other clade, but this alone will not ensure the mutual exclusivity implied by the name *Parahebe*. Mutual exclusivity can be ensured by using the type of each name as an external specifier for the other name, or by including a qualifying clause making the name *Parahebe* inapplicable in the context of any phylogeny in which the two clades are not mutually exclusive. However, neither of these approaches should be taken if the accepted usage (at the time when the definition is prepared) treats *Hebe* and *Parahebe* as nested.

Note 11G.1. The following prefixes and suffixes imply greater inclusiveness than the base name: *Holo-, Pan-, -formes, -morpha*. The following prefixes imply lesser inclusiveness than the base name: *Eo-, Eu-, Neo-, Proto-*. The following prefixes imply mutual exclusivity with the base name: *Pseudo-, Para-*. These are not intended to be exhaustive lists.

Note 11G.2. The creation of new clade names that add such a prefix to a preexisting or converted name with a rank-specific ending may lead to confusion for users of the rank-based system because the new name may be incorrectly taken to imply the existence of a genus name with the same prefix.

Example 1. The creation of a new clade name *Protobrassicaceae* from the base name *Brassicaceae* might be incorrectly taken to imply the existence of a genus *Protobrassica*.

Recommendation 11H. When defining the names of low-level clades that coincide with or overlap the boundaries of species, differences in species criteria and hypothesized species boundaries may result in a phylogenetically defined name being applied to different clades even in the context of the same phylogeny. Recommendations 11H and I are intended to reduce the likelihood of this undesirable outcome and address it when it occurs. If the definitional author is aware that using a particular species as a specifier may result in the application of the name to a different clade than if the type specimen of that species' name were instead used, one or the other should be unambiguously selected as the specifier (see Arts. 11.4 and 11.6 for the ramifications of that choice), and the situation should be clearly explained in the protologue.

Recommendation 11I. If, after establishment of a clade name, it is discovered that a species or type specimen used as a specifier results in the name being applied to a different clade (in the context of the same phylogeny) than if the other entity, i.e., the type specimen or species, had been used instead, and if the ambiguity would be eliminated by selecting the other entity as the specifier, then an unrestricted emendation (Arts. 15.11–15.13) designating that entity as the specifier may be published (see also Rec. 11J). It is preferable, though not required, that the emendation be published by the original definitional author(s) (Art. 15.14). If such an emendation is published by anyone other than the original definitional author(s), the intent of the original author(s) should be considered carefully and addressed in the protologue of the emendation (see Note 15.11.4 and Arts. 15.12 and 15.13). If the protologue of the original definition included a discussion of the choice of species versus type specimens as specifiers (see Rec. 11H), this should be viewed as part of the author(s)' intent. Specifically, if the author(s) stated that their preference for using type specimens (or conversely, species) as specifiers regardless of the effect on clade composition, that preference should be honored; in such a case, an unrestricted emendation by anyone other than the author(s) would be inappropriate.

Recommendation 11J. If an emendation is published in the situation described in Recommendation 11I, and if a species originally used as a specifier has more than one type (i.e., syntypes), whichever syntype is chosen as the new specifier should be simultaneously designated as the lectotype of that species under the appropriate rank-based code.

Chapter V

Selection of Accepted Clade Names

ARTICLE 12. PRECEDENCE

12.1. Nomenclatural uniqueness is achieved through precedence, the order of preference among established names. When homonyms or synonyms exist, precedence determines the selection of accepted names.

Note 12.1.1. Although the entity to which precedence applies in this code is referred to as a name, it is really the combination of a name and its definition. In different cases, one or the other of these components is more important. Specifically, in the case of synonyms, precedence refers primarily to the name, whereas in the case of homonyms, precedence refers primarily to the definition.

12.2. Precedence is based on the date of establishment, with earlier-established names having precedence over later ones, except that later-established names may be conserved over earlier ones under the conditions specified in Article 15, and panclade names (Art. 10.3) have precedence under the conditions specified in Article 14.4.

Note 12.2.1. In the case of homonymy involving names governed by two or more rank-based codes (e.g., the application of the same name to a group of animals and a group of plants), precedence is based on the date of establishment under this code. However, the Committee on Phylogenetic Nomenclature (see Art. 22) has the power to conserve a later-established homonym over an earlier-established homonym. This might be done if the later homonym is much more widely known than the earlier one.

12.3. For the determination of precedence, the date of establishment is considered to be the date of publication (see Art. 5), not the date of registration (but see Arts. 13.4 and 14.3).

ARTICLE 13. HOMONYMY

13.1. Homonyms are names that are spelled identically but refer to different taxa. In this code, all homonyms are established and identically spelled clade names based on different phylogenetic definitions. However, not all identically spelled clade names based on different phylogenetic definitions are necessarily homonyms because different definitions may refer to the same clade under some phylogenetic hypotheses but not under others.

Example 1. Suppose that Pedersen defined *Lamiaceae* as the name of the smallest clade containing *Lamium album* Linnaeus 1753 and *Congea tomentosa* Roxburgh 1819, and Ramírez defined *Lamiaceae* as the name of the smallest clade containing *Lamium album* Linnaeus 1753 and *Symphorema involucratum* Roxburgh 1798. If so, these two definitions would refer to the same clade in the context of any phylogeny in which *Congea tomentosa* and *Symphorema involucratum* share a more recent common ancestor with each other than either does with *Lamium album* but not if *Congea tomentosa* shares a more recent common ancestor with *Lamium album* than it does with *Symphorema involucratum*.

Note 13.1.1. Homonyms result when an author establishes a name that is spelled identically to, but defined differently than, an earlier established name. This situation can occur either when an author is unaware of the earlier establishment of an identically spelled but differently defined name (Ex. 1) or when an author knowingly adopts an earlier established name but proposes, either deliberately or inadvertently, a different definition for that name (Ex. 2). Although names in the second scenario can be considered the same name in the sense that one use is derived from the other (see Note 9.15A.2), the identically spelled names in both scenarios are treated as homonyms under this code because they have different definitions.

Example 1. If Mukherjee defined *Prunella* as the name of the smallest clade containing *Prunella modularis* Linnaeus 1758 and *Prunella collaris* Scopoli 1769 (which are birds), and Larsen defined *Prunella* as the name of the smallest clade containing *Prunella laciniata* Linnaeus 1763, *Prunella grandiflora* Scholler 1775, *Prunella vulgaris* Linnaeus 1753, and *Prunella hyssopifolia* Linnaeus 1753 (which are plants), *Prunella* of Mukherjee and *Prunella* of Larsen would be homonyms.

Example 2. Gauthier et al. (1988) defined the name *Lepidosauromorpha* as referring to the clade composed of *Lepidosauria* and all organisms sharing a more recent common ancestor with *Lepidosauria* than with *Archosauria* (a maximum-clade definition). Laurin (1991) defined the name *Lepidosauromorpha* as referring to the clade originating with the most recent common ancestor of *Palaeagama*, *Saurosternon*, *Paliguana*, *Kuehneosaurus*, and *Lepidosauria* (a minimum-clade definition). If this code had been in effect when these names were published, *Lepidosauromorpha* of Gauthier et al. and *Lepidosauromorpha* of Laurin would have been homonyms.

13.2. Phylogenetic definitions are considered to be different if either: (1) they are of the same kind (e.g., minimum-clade) but cite different specifiers and/or have different restrictions specified in their qualifying clauses (if any), or (2) they are of a different kind (e.g., minimum-clade vs. maximum-clade).

Note 13.2.1. Alternative wordings of minimum-clade definitions such as those provided in Article 9.5 are not considered to be different, provided they are based on the same specifiers and have the same restrictions (Arts. 11.12–11.14). The same is true of alternative wordings of maximum-clade definitions (Art. 9.6), apomorphy-based definitions (Art. 9.7), directly-specified-ancestor definitions (Note 9.5.1), minimum-crown-clade definitions (Art. 9.9), maximum-crown-clade definitions (Art. 9.9), apomorphy-modified crown-clade definitions (Art. 9.9), maximum-total-clade definitions (Art. 9.10), crown-based total-clade definitions (Art. 9.10), and other types of phylogenetic definitions that are not explicitly mentioned in this code.

13.3. If two or more definitions have been established for identically spelled clade names, the only acceptable name (i.e., the combination of name and definition; see Note 12.1.1) is the first one established under this code. A later homonym, unless conserved, is not an acceptable name of any clade.

13.4. When two or more homonyms have the same publication date (Art. 5), the one that was registered first (and therefore has the lowest registration number) takes precedence.

13.5. If the oldest name of a clade is not acceptable because it is a later homonym, it is to be replaced by the established name that has precedence. If all established names that apply to the clade are not acceptable because they are later homonyms, a replacement name may be explicitly substituted for the earliest-established name that applies to the clade. A replacement name must be established, following the procedures in Articles 7, 13.6, and 13.7. The definition of a replacement name for a clade is the definition of the name it replaces.

13.6. In order to be established, a replacement name must be clearly identified as such in the protologue where the replacement is published, by the designation "replacement name" or "*nomen substitutum.*"

13.7. In order for a replacement name to be established, the replaced name on which it is based must be clearly indicated by a direct and unambiguous bibliographic citation (see Art. 9.16) that includes its author, date, and the journal or book in which the name was originally published. The registration number of the replaced name must also be cited.

ARTICLE 14. SYNONYMY

14.1. Synonyms are names that are spelled differently but refer to the same taxon. In this code, synonyms must be established clade names and may be homodefinitional (based on the same definition) or heterodefinitional (based on different definitions). The criteria for determining whether definitions are different are described in Article 13.2, including Note 13.2.1.

Note 14.1.1. Homodefinitional synonyms are synonyms regardless of the phylogenetic context in which the names are applied. However, in the case of names with different definitions, the phylogenetic context determines whether the names are heterodefinitional synonyms or not synonymous.

Example 1. Suppose that *Hypothetica* were defined as the smallest clade containing species A and B, and *Cladia* were defined as the smallest clade containing species C and B. In the context of any hypothesized phylogeny in which A shares a more recent common ancestor with C than either does with B, *Hypothetica* and *Cladia* would be heterodefinitional synonyms. However, in the context of an alternative hypothesis that A and B are more closely related to each other than either is to C, *Hypothetica* and *Cladia* would not be synonymous.

Note 14.1.2. Minimum-clade, apomorphy-based, and maximum-clade definitions (Arts. 9.5–9.7) usually designate different clades, although they may be nested clades that differ only slightly in inclusiveness. Therefore, names based on two or more of these different kinds of definitions usually are not synonyms. (In theory, it is possible for different types of definitions to designate the same clade. For example, in cases in which doubling of the chromosomes (auto-polyploidy) causes speciation, the apomorphic chromosome number arises simultaneously with the splitting of a lineage. In such cases, an apomorphy-based definition that uses this chromosome number as a specifier will refer to the same clade as a maximum-clade definition that uses the species in which the chromosome doubling occurred, or one of its descendants, as the internal specifier if an appropriate external specifier is used.)

14.2. If there are two or more synonyms for a clade, the accepted name for that clade is the earliest acceptable one that applies to it, except in cases of conservation (Art. 15), or precedence of a panclade name (Art. 14.4), or precedence of a younger panclade name over an older one to maintain consistency with crown-clade names (Art. 14.5).

14.3. When two or more synonyms have the same publication date (Art. 5), the one that was registered first (and therefore has the lowest registration number) takes precedence.

14.4. If a panclade name (Art. 10.3) and a name that was not explicitly established as applying to a total clade are judged to be hetero-definitional synonyms (Art. 14.1), the panclade name has precedence even if it was established later (except in cases covered by Art. 10.7).

14.5. In order to maintain the relationships between panclade names and the crown-clade names upon which they are based, precedence among panclade names is based on precedence of the corresponding crown-clade names. Thus, if two or more panclade names are considered synonyms because the names of the crown clade upon which they are based are considered synonyms, the panclade name that has precedence is the one that is based on the crown-clade name that has precedence, and that is the case even if one or more of the other panclade names were established earlier than the one based on the crown-clade name that has precedence.

Example 1. Suppose that first the crown-clade name *Reptilia* was established, then later the crown-clade name *Lepidosauria* and the panclade name *Pan-Lepidosauria* were established simultaneously, then later still the panclade name *Pan-Reptilia* was established. In the context of phylogenies in which *Reptilia* and *Lepidosauria* are synonyms, *Reptilia* would be the accepted name of the crown clade, and *Pan-Reptilia* would be the accepted name of the corresponding total clade even though *Pan-Lepidosauria* was established before *Pan-Reptilia*.

ARTICLE 15. CONSERVATION, SUPPRESSION, AND EMENDATION

15.1. Conservation of names and emendation of definitions are means of overriding precedence based on date of establishment (Art. 12.2) in the interest of stability (e.g., in terms of composition and/or diagnostic characters).

15.2. Conservation of names is possible only under extraordinary circumstances and requires approval of the Committee on Phylogenetic Nomenclature (CPN; see Art. 22).

15.3. Once a name has been conserved, the entry for the affected name in the registration database is to be annotated to indicate its conserved status relative to other names that are simultaneously suppressed. The entries for suppressed names are to be similarly annotated.

15.4. In the case of heterodefinitional synonyms, the earlier name may be conditionally suppressed so that it may be used when not considered synonymous with the later name. In the case of homonyms and homodefinitional synonyms, suppression is unconditional.

15.5. When a name is unconditionally suppressed, there are no conditions under which it has precedence with regard to either synonymy or homonymy. Therefore, if a homodefinitional synonym has been suppressed, that name can be established subsequently with a different definition as an acceptable name.

15.6. When a conserved name competes with names against which it has not been explicitly conserved, the earliest established of the competing names has precedence.

15.7. Although names are normally suppressed only when a synonym or homonym is conserved, the CPN may unconditionally suppress a name if it is nomenclaturally disruptive, without necessarily conserving an alternative. An unconditionally suppressed name can be established subsequently with a different definition as an acceptable name.

15.8. An emendation is a formal change in a phylogenetic definition. A restricted emendation requires approval by the CPN (see Art. 22), while an unrestricted emendation may be published without CPN approval.

15.9. All emendations must be published (Art. 4) and registered (Art. 8).

15.10. A restricted emendation (see Art. 15.8) is intended to change the application of a particular name through a change in the conceptualization of the clade to which it refers. Restricted emendations may involve changes in definition type, clade category, specifiers, and/or qualifying clauses.

Note 15.10.1. A restricted emendation is a mechanism to correct a definition that fails to associate a name with the clade to which it has traditionally referred, even in the context of the reference phylogeny adopted by the original definitional author.

Example 1. Suppose the name *Angiospermae* had been defined as the smallest clade containing *Zea mays* Linnaeus 1753 and *Gnetum gnemon* Linnaeus 1767. By including *Gnetum*, this definition specifies a more inclusive clade than the one to which the name *Angiospermae* traditionally refers. Correcting the definition would qualify as a restricted emendation (i.e., it would require approval by the CPN).

15.11. An unrestricted emendation (see Art. 15.8) is intended to preserve the application of a particular name in terms of the conceptualization of the clade to which it refers. Unrestricted emendations may involve changes in specifiers or qualifying clauses, or clarification of the meaning of "extant," but must retain the same clade category (i.e., crown clade, total clade, or neither) if category was specified in the protologue, the same definition type (minimum-clade, maximum-clade, or apomorphy-based) except as specified in Note 15.11.2, and the same clade conceptualization as interpreted from the protologue.

Note 15.11.1. An unrestricted emendation is a mechanism to prevent undesirable changes in the application of a particular name (in terms of clade conceptualization) when the original definition is applied in the context of a revised phylogeny.

Example 1. Several recent phylogenetic analyses suggest that *Amborella trichopoda* is sister to the rest of *Angiospermae*, but evidence for this position of *Amborella* was not discovered until the late 1990s. If, prior to this discovery, *Angiospermae* had been given a minimum-crown-clade definition that did not include *Amborella trichopoda* as an internal specifier, *Angiospermae* would not include *Amborella* according to its currently inferred relationships. However, the definitional author would presumably have intended for *Amborella* to be included in *Angiospermae* because it has always been included in the taxon designated by that name. In such a situation, an unrestricted emendation that adds *Amborella trichopoda* to the list of internal specifiers would avoid an undesirable change in clade composition and would be consistent with the clade conceptualization of the original definitional author and with historical use.

Note 15.11.2. In the context of this article, minimum-crown-clade, maximum-crown-clade, and apomorphy-modified crown-clade definitions are all considered the same definition type, so it is permissible for an unrestricted emendation to change from one to another of these three variants of the crown-clade definition, provided that all internal specifiers are extant.

Example 1. In the situation described in Note 15.11.1, Example 1, as an alternative to adding *Amborella trichopoda* to the list of specifiers, it might be preferable to change the original minimum-crown-clade definition to a maximum-crown-clade definition such as "the largest crown clade containing *Zea mays* Linnaeus 1753 but not *Cycas circinalis* Linnaeus 1753, *Gnetum gnemon* Linnaeus 1767, *Ginkgo biloba* Linnaeus 1771, and *Pinus sylvestris* Linnaeus 1753." Such a definition avoids the need for further emendation if some other species (i.e., other than *Amborella trichopoda*) or subclade is inferred in the future to be sister to the rest of the angiosperms.

Note 15.11.3. If it is specified in the protologue that the name refers to a crown clade or a total clade, this clade category cannot be changed through an unrestricted emendation. If the clade category is not specified in the protologue, the category may still play a role in determining the author's conceptualization of the clade (see Note 15.11.4 and Art. 15.13, Ex. 1). The category of crown clade is considered to be specified in the protologue if any of the specified formulations of crown-clade definitions or their standard abbreviations (see Art. 9.9) is used. The category of total clade is considered to be specified in the protologue if a panclade name (Art. 10.3) or any of the formulations of total-clade definitions or their standard abbreviations (Art. 9.10) is used.

Note 15.11.4. Interpretation of the original definitional author's clade conceptualization is based on the definition and all other information in the protologue. Important components of the definitional author's conceptualization of the clade include (but are not necessarily restricted to) composition, apomorphies, clade category (e.g., crown versus non-crown; see Art. 15.13, Ex. 1), the existence of a basal dichotomy into two particular subclades (see Art. 15.13, Ex. 2), and conceptualization of a clade as an entire branch regardless of composition (see Art. 15.13, Ex. 3).

15.12. The protologue of an unrestricted emendation must provide evidence that the conceptualization of the clade is the same as that of the original definitional author. The protologue must also explain why the emended definition is preferable to the definition being emended.

15.13. If conflicting evidence from the protologue (see Note 15.11.4) makes it unclear whether a proposed emendation is consistent with the original conceptualization of the clade, the emendation must be considered by the CPN (i.e., it must be a restricted emendation). Disagreements within the systematics community as to whether a published unrestricted emendation changes the conceptualization of a clade (i.e., whether the emendation should have been restricted) are to be resolved by referring the issue to the CPN for a decision (see Art. 22).

Example 1. Suppose that the name *Mammalia* had first been defined phylogenetically as the smallest clade containing *Ornithorhynchus anatinus* (Shaw 1799) and *Homo sapiens* Linnaeus 1758, both of which are extant, without explicitly stating that the name refers to a crown clade. Further, suppose that the definitional author(s) had considered *Mammalia* to include *Multituberculata* (a wholly extinct group). If, under a newly proposed phylogenetic hypothesis, *Multituberculata* is no longer included in *Mammalia* under the stated definition, and if the definitional author(s) did not indicate whether reference to a crown clade or inclusion of *Multituberculata* was more fundamental to their use of the name *Mammalia*, conflicting evidence exists concerning the original conceptualization of that taxon. Therefore, adding a member of *Multituberculata* to the set of internal specifiers in the definition of *Mammalia*, or otherwise modifying the definition of *Mammalia* so that it refers to a clade that includes *Multituberculata*, would require consideration by the CPN.

Example 2. Suppose that the name *Dinosauria* had first been defined phylogenetically as the smallest clade containing *Megalosaurus bucklandii* von Meyer 1832 (*Saurischia*) and *Iguanodon bernissartensis* Boulenger in Beneden 1881 (*Ornithischia*). Further, suppose that the definitional author(s) had considered *Dinosauria* to include *Herrerasauridae*. If, under a newly proposed phylogenetic hypothesis, *Herrerasauridae* is no longer included in *Dinosauria* under the stated definition, and if the definitional author(s) did not indicate whether inclusion of *Herrerasauridae* or application to the clade whose basal dichotomy is represented by *Saurischia* and *Ornithischia* was more fundamental to their use of the name *Dinosauria*, conflicting evidence exists concerning the original conceptualization of that taxon. Therefore, adding a species of *Herrerasauridae* to the set of internal specifiers in the definition of *Dinosauria*, or otherwise modifying the definition of *Dinosauria* so that it refers to a clade that includes *Herrerasauridae*, would require consideration by the CPN.

Example 3. Suppose that the name *Saurischia* had first been defined phylogenetically as referring to the largest clade containing *Allosaurus fragilis* Marsh 1877 but not *Stegosaurus armatus* Marsh 1877 (*Ornithischia*). Further, suppose that the definitional author(s) had considered *Saurischia* to include *Herrerasauridae*. If, under a newly proposed phylogenetic hypothesis, *Herrerasauridae* is no longer included in *Saurischia* under the stated definition, and if the definitional author(s) did not indicate whether inclusion of *Herrerasauridae* or application to the sister clade of *Ornithischia* was more fundamental to their use of the name *Saurischia*, conflicting evidence exists concerning the original conceptualization of that taxon. Therefore, adding a species of *Herrerasauridae* to the set of internal specifiers in the definition of *Saurischia*, or otherwise modifying the definition of *Saurischia* so that it refers to a clade that includes *Herrerasauridae*, would require consideration by the CPN.

15.14. Although anyone may publish an unrestricted emendation, it is preferable that the emendation be authored or coauthored by the author or authors of the original definition. If one or more of the original definitional authors are still alive, another worker who thinks that an unrestricted emendation is warranted must provide evidence when registering the emendation that the first author of the original definition (or the second, third authors, etc., if the first author is deceased or otherwise unable to respond) was contacted and offered the opportunity to co-author the emendation.

Note 15.14.1. Minimal evidence required for registration of an unrestricted emendation includes the e-mail address or phone number of the original definitional author(s) contacted and the date when the contact was made. If all of the original definitional authors are deceased or otherwise unable to respond, this information must be submitted to the registration database as well. Supplementary information such as the text of the definitional author(s)' response may also be submitted.

Note 15.14.2. Although the author or authors of the original definition must be offered the opportunity to co-author an emendation of the original definition, it is not necessary that they be offered the opportunity to co-author the entire publication in which the emendation appears (see Art. 19.2).

15.15. Within a phylogenetic context in which the original definition and an unrestricted emendation apply to the same clade, the original definition has precedence.

Chapter VI

Hybrids

ARTICLE 16. PROVISIONS FOR HYBRIDS

16.1. Hybrid origin of a clade may be indicated by placing the multiplication sign (×) in front of the name. The names of clades of hybrid origin otherwise follow the same rules as for other clades.

16.2. An organism that is a hybrid between named clades may be indicated by placing the multiplication sign between the names of the clades; the whole expression is then called a hybrid formula.

Recommendation 16.2A. In cases in which it is not clear whether a set of hybrid organisms represents a clade (as opposed to independently produced hybrid individuals that do not form a clade), authors should consider whether a name is really needed, bearing in mind that formulae, though more cumbersome, are more informative.

Chapter VII

Orthography

ARTICLE 17. ORTHOGRAPHIC REQUIREMENTS FOR ESTABLISHMENT

17.1. In order to be established, a clade name must be a single word and begin with a capital letter. The name must be composed of more than one letter and consist exclusively of letters of the Latin alphabet as used in contemporary English, which is taken to include the 26 letters a, b, c, d, e, f, g, h, i, j, k, l, m, n, o, p, q, r, s, t, u, v, w, x, y, and z, even though some of these letters are rare or absent in classical Latin. If other letters, ligatures, numerals, apostrophes, or diacritical signs that are foreign to classical Latin appear in a name, it cannot be established. A hyphen may be included in a clade name only when it is a panclade name (see Art. 10.3), or the name has an apomorphy-based definition and is formed in accordance with Article 10.8, or the name is based on the preexisting name of a subdivision of a genus (see Rec. 10F), or the name is based on a preexisting name preceded by a taxon-related prefix such as *Phyto-*, *Phyco-*, *Myco-*, *Prokaryo-*, or *Zoo-* in the situation covered by Recommendation 10D. When other letters, ligatures, or diacritical signs appear in the protologue of a preexisting name, they must be transliterated at the time of conversion in conformity with the rank-based code that is applicable to the clade concerned. Hyphens or apostrophes present in a preexisting name must be deleted at the time of conversion. See Note 18.1.2 for the inclusion of diaereses and apostrophes as optional pronunciation guides in the subsequent use of established names.

17.2. When a preexisting name has been published in a work where the letters u and v or i and j are used interchangeably, or are used in any other way incompatible with modern practices (e.g., one of those letters is not used or is used only when capitalized), those letters must be transliterated at the time of conversion in conformity with modern usage.

Example 1. *Vffenbachia* Fabr. (1763) would be changed to *Uffenbachia* when converted.

17.3. A clade name may be a word in or derived from Latin, Greek, or any other language provided that the name uses the Latin alphabet (Art. 17.1).

Recommendation 17.3A. If a clade name is derived from a language other than Latin, it should be Latinized, in the tradition of scientific names governed by the *ICNAFP*, *ICZN*, etc.

Recommendation 17.3B. In order to avoid confusion with vernacular and informal names, a new clade name should not be spelled identically to a vernacular or informal name in any modern language. However, the scientific name may be derived from the vernacular or informal name by Latinization.

Example 1. "Tricolpates" (a plant clade) is an informal name and should therefore not be adopted as the formal scientific name for this (or any other) clade. However, a name derived by Latinizing "tricolpates" (e.g., *Tricolpatae*) may be used.

17.4. If a clade is named after a person, the clade name, in order to be established, must differ in spelling from the person's name, for example through the addition of a Latinized ending.

Example 1. If a clade is named in honor of a person whose surname is Woodson, the clade name must not be *Woodson* but may be *Woodsonia*.

17.5. In order to be established, the spelling of a converted name must be identical to that of the preexisting name on which it is based, except as noted in Articles 17.1 and 17.2.

Recommendation 17.5A. When a preexisting name is converted, the spelling in prevailing use should be retained. As a general guideline, adoption of a spelling by two-thirds of the authors who have used the name in the past 25 years would qualify as prevailing use. If it is not clear which spelling is the prevailing one, the original spelling should be adopted for the converted name, except for the correction of orthographical (including typographical) errors and the mandatory corrections imposed under Articles 17.1 and 17.2. In this code, the original spelling is the one used in the protologue.

Recommendation 17A. Names established under this code should be pronounceable. Thus, every syllable should contain a vowel (or diphthong), and combinations of consonants that do not generally occur in either Latin or English should be avoided unless they are contained within the name of a person, place, or other entity after which a clade is named.

Recommendation 17B. New clade names should follow the rules and recommendations of the appropriate rank-based code with regard to Latin grammar. However, failure to follow those rules and recommendations does not invalidate the establishment of names under this code.

ARTICLE 18. SUBSEQUENT USE AND CORRECTION OF ESTABLISHED NAMES

18.1. The original spelling of a name established under this code is the correct spelling and should be retained in subsequent publications, except for the correction of typographical errors (see Art. 18.5). The original spelling is the one that is used in the protologue at the time of establishment and that is registered (see Art. 8).

Note 18.1.1. The original spelling of a converted name is correct so long as it is based on one of the spellings of the preexisting name, even if the prevailing spelling was not adopted (see Rec. 17.5A).

Note 18.1.2. Use of a diaeresis to indicate that a vowel is to be pronounced separately from the preceding vowel is not part of the spelling (orthography) of a name, but it may be included in an established name as an optional pronunciation guide. Similarly, use of an apostrophe to indicate a break between syllables is not part of the spelling of a name, but it may be included in an established name as an optional pronunciation guide.

18.2. Spellings that do not follow Recommendation 17B (e.g., incorrect Latinization or use of an inappropriate connecting vowel) and spellings that contain incorrect transliterations are not to be corrected.

18.3. If the registered spelling of a name disagrees with the spelling in the protologue or the name is spelled more than one way in the protologue, the author should determine which is correct and notify the registration database administrator promptly.

Note 18.3.1. If the author notifies the database administrator that the registered spelling is incorrect, the administrator will correct the database and insert a note that the change was made. If one or more spellings in the protologue are incorrect, the administrator will annotate the database to alert users that this is the case.

18.4. If the registered spelling of a name disagrees with the spelling in the protologue or the name is spelled more than one way in the protologue, and the author is no longer alive or is otherwise unable to determine which spelling is correct, the following guidelines are to be used: If it is clear that all but one of the spellings are typographical errors, the remaining one is treated as correct. If it is not clear which spellings are typographical errors, the one that is most consistent with Recommendation 17B is treated as correct. If it is not clear which spellings are typographical errors, and it is not clear that one is more consistent with Recommendation 17B than the others, the one immediately associated with the designation "new clade name," "converted clade name," etc. is treated as correct. Such decisions regarding the correct spelling of a name if made by anyone other than the author, must be published (Art. 4) before the registration database administrator is notified (see Rec. 18A).

Note 18.4.1. If the author of a published correction notifies the database administrator that the registered spelling is incorrect, the administrator will correct the database and insert a note that the change was made. If one or more spellings in the protologue are incorrect, the administrator will annotate the database to alert users that this is the case.

Recommendation 18.4A. The person making an orthographic correction of the sort covered by Article 18.4 should notify the database administrator promptly after publishing it.

18.5. If the registered spelling of a name and the spelling in the protologue agree but contain a typographical error, the author may publish a correction. If the author is no longer alive or is otherwise unable to correct the error, any person may publish a correction (see Rec. 18A).

Note 18.5.1. After the registration database administrator is notified, the spelling will be corrected in the database and a note will be added stating that the change was made.

Note 18.5.2. A correction slip inserted in the original publication does not qualify as a published correction. Publication of corrections must satisfy the requirements of Article 4.

18.6. Accidental misspellings of a name that appear in print subsequent to establishment are not to be treated as new names but as incorrect spellings of the established name. The same is true of unjustified corrections (i.e., any correction that does not fall under Arts. 18.3–18.5, particularly those that violate Art. 18.2).

Recommendation 18A. The person making an orthographic correction of the sort covered by Articles 18.4 and 18.5 should notify the database administrator promptly after publishing it.

Chapter VIII

Authorship

ARTICLE 19. AUTHORSHIP OF NAMES AND DEFINITIONS

19.1. The nominal author(s) of a clade name is (are) the person(s) who first published the name (but see Notes 9.15A.1, 19.1.1), regardless of whether it was phylogenetically defined and regardless of whether it was initially applied to a taxon that differed somewhat in composition from the clade for which the name is being converted (provided that it is not a homonym; see Note 9.15A.2). The definitional author(s) of a clade name is (are) the person(s) who established that name, including publication of a phylogenetic definition for it (either the original definition or an emended one), under this code.

Note 19.1.1. When determining nominal authorship, publications before the nomenclatural starting point for the particular group of organisms under the appropriate rank-based code are not considered.

Note 19.1.2. For a new clade name (except a new replacement name), the nominal and definitional authors are the same. For a converted clade name or a replacement name, the nominal and definitional authors are frequently different.

Note 19.1.3. The *ICNAFP* (Art. 39) requires that new taxon names (other than for fossils) published from 1935 through 2011 be accompanied by a Latin description or diagnosis or a reference to the same. By contrast, this code does not require a Latin description or diagnosis for the establishment of a name. Therefore, a person who published a preexisting botanical name with a description or diagnosis and otherwise satisfying the *ICNAFP* rules for a legitimate name (see Art. 6.2) may be considered the nominal author, regardless of whether the description or diagnosis was in Latin (although there are situations in which such a person might not be considered the nominal author; see Note 9.15A.1, Ex. 1).

19.2. A clade name or definition is to be attributed to the author(s) of the protologue, even though authorship of the publication as a whole may be different.

Note 19.2.1. In some cases, a breadth of evidence may need to be considered to determine the correct author attribution, including ascription of the name, statements in the introduction, title, or acknowledgements, typographical distinctions in the text, and even statements made in other volumes and editions in the same series or in entirely different publications.

Note 19.2.2. In the absence of evidence to the contrary, the authorship of the protologue can be assumed to be the same as the authorship of the entire publication in which it appears.

19.3. The nominal authors of a replacement name are the authors of that name, not the authors of the replaced name. However, because the definition remains the same (Art. 13.5), the definitional authors of the replacement name are the definitional authors of the replaced name.

19.4. A preexisting clade name is to be attributed to the author(s) of the protologue when only the name, but not the rest of the protologue, is attributable to (a) different author(s) (see Art. 20.8).

19.5. When the prevailing spelling of a preexisting name differs from the original spelling due to correction of orthographic or typographical errors in the original spelling or orthographic standardizations, whether imposed by a rank-based code or accepted by convention, the prevailing spelling is to be attributed to the author of the publication in which the original spelling was used.

Example 1. *Iguana*, which is the prevailing spelling of the name, is attributed to Linnaeus even though he used the spelling *Igvana* in the original publication.

Note 19.5.1. Article 19.5 does not apply to names whose spellings have been "corrected" under a rank-based code to the standard ending for the rank at which it was published (see Note 9.15A.4).

Example 1. Under this code, the name *Hypericaceae* is not attributed to Jussieu, who published *Hyperica* as a family name, even though *Hypericaceae* is considered to be a "correction" of *Hyperica* under the *ICNAFP*, i.e., through addition of the standard ending for plant families (see Note 9.15A.4, Ex. 1).

Chapter IX

Citation

ARTICLE 20. CITATION OF AUTHORS AND REGISTRATION NUMBERS

20.1. Citation of nominal and definitional authors (Art. 19.1) is optional, but if authors are cited, Articles 20.2–20.8 are to be followed.

20.2. Authors' names are to be cited after the clade name. Nominal authors of any name, whether preexisting or new, are to be cited without enclosing symbols. Definitional authors, if different from the nominal author(s) of the name in question (Note 19.1.2 and Note 20.2.1), are to be cited within enclosing symbols: the authors of the original definition (i.e., the definitional authors of a converted name) are to be cited in square brackets ([ ]; Art. 20.4, Ex. 1); the authors of an emended definition are to be cited in braces ({ }; Art. 20.6, Ex. 1).

Note 20.2.1. If the nominal and definitional authors are the same (as for most new clade names; see Note 19.1.2), they are to be cited as nominal authors, and thus no enclosing symbols are to be used.

20.3. If more than one set of authors is cited, they are to be cited in the following order: nominal author(s) of the preexisting or new name (including a replacement name); author(s) of the original definition; author(s) of an emended definition.

20.4. If the definitional authors of a converted name are cited, the nominal authors of the preexisting name on which it is based, if known, must also be cited.

Example 1. Suppose that Larson established a converted clade name *Hypotheticus* in 2020 based on the preexisting name *Hypotheticus* of Meekins (published in 1956). In this situation, the citation of the converted name would be *Hypotheticus* Meekins [Larson]. Meekins is the nominal author; Larson is the definitional author.

Example 2. If Larson established a converted clade name *Hypotheticus* based on the preexisting name *Hypotheticus*, and if the authorship of this name were unknown, the citation of the converted name would be *Hypotheticus* [Larson].

Note 20.4.1. The publication years of the preexisting name and converted name may follow the names of the respective authors.

Example 1. Using Example 1 of Article 20.4, the citation with publication years would be *Hypotheticus* Meekins 1956 [Larson 2020].

Recommendation 20.4A. If a preexisting name was used in association with more than one rank or composition, and authorship is cited, the nominal author(s) cited should be the original author(s) of the name, as spelled for the purpose of conversion, rather than the first author(s) who applied the name later in association with a different rank or composition (but see Rec. 9.15A).

Recommendation 20.4B. If a preexisting name has been attributed to an author or authors other than the first author(s) who used the name being converted (as can occur under the Principle of Coordination of the *ICZN*), the nominal author(s) cited should not be the former but rather the author(s) of the name as spelled for the purpose of conversion (but see Note 9.15A.3 and its Ex. 1).

20.5. If the nominal authors of a replacement name are cited, the definitional authors of the replacement name (i.e., the definitional authors of the replaced name; see Arts. 13.5, 19.3) must also be cited.

Example 1. Suppose that Holmes was the definitional author of the name *Cladus*, which turned out to be a later homonym of *Cladus* (established by a different author), and then Clarke published the new name *Imaginarius* as a replacement name for *Cladus* Holmes. The full citation of the replacement name would be *Imaginarius* Clarke [Holmes]. If, instead, Clarke had converted the replacement name from the preexisting name *Fabricatus* Merriam, the full citation of the replacement name would be *Fabricatus* Merriam [Holmes].

20.6. If the authors of an emended definition (see Art. 15) are cited, the authors of the original definition must also be cited.

Example 1. If *Fictitius* was established as a new name by Stein, and Maki subsequently emended Stein's definition, the full citation would be *Fictitius* Stein {Maki}. If, instead, Stein had converted *Fictitius* from the preexisting name *Fictitius* Merriam, the full citation would be *Fictitius* Merriam [Stein] {Maki}.

20.7. When authorship of a name differs from authorship of the publication in which it is established, both may be cited, connected by the word "in." In such a case, "in" and what follows are part of a bibliographic citation and are only to be included if the publication is referred to, at least by its year.

20.8. The optional use of "ex" under the *ICNAFP* to cite author(s) to whom the name, but not the rest of the protologue, is attributable is not adopted in this code.

Recommendation 20A. Bibliographic references to the protologue of established names are available in the registration database and may be accessed by either clade name or registration number. However, only the registration number is reliably unique. Therefore, in cases of potential ambiguity, the registration number should be cited at least once in any publication in which the corresponding name is used.

Chapter X

Species Names

ARTICLE 21. PROVISIONS FOR SPECIES NAMES

21.1. This code does not govern the establishment or precedence of species names or names associated with ranks below species under the rank-based codes (e.g., *ICNP*, *ICNAFP*, *ICZN*). To be considered available (*ICZN*) or validly published (*ICNAFP*, *ICNP*), the name of a species or infraspecific taxon must satisfy the provisions of the appropriate rank-based code. This article contains recommendations about how to publish or use previously published names of species and infraspecific taxa governed by rank-based codes in conjunction with clade names governed by this code.

Note 21.1.1. In Article 21.1, the term "names of species and infraspecific taxa" does not refer to names established under this code that apply to clades that correspond in composition to or are nested within taxa that are ranked as species under a rank-based code.

21.2. The name of a species under the rank-based codes (except the *ICVCN*) is a binomen (two-part name), the first part of which is a generic name (i.e., a name that is tied to the rank of genus) and the second part of which is a specific name (*ICZN*) or epithet (*ICNP*, *ICNAFP*) (i.e., a name that is tied to the rank of species). Because this code is independent of categorical ranks (Art. 3.1), the first part of a species binomen is not interpreted as a genus name but simply as the name of a taxon that includes that species.

21.3. To satisfy the requirements of the rank-based codes (see Art. 21.2), a specific or infraspecific name (*ICZN*) or epithet (*ICNAFP*) must be published in unambiguous combination with a name that is implicitly or explicitly associated with the rank of genus (even though it may not have been established as a clade name under this code). For names governed by the *ICZN*, this practice must be followed throughout the publication that establishes the name (*ICZN* Art. 11.4).

Recommendation 21.3A. When establishing a new species name (binomen) under the appropriate rank-based code, some mechanism should be used to indicate whether the genus name is an established clade name under this code. If symbols are used, their meaning should be made clear.

Example 1. [P]*Hypotheticus* could indicate that *Hypotheticus* is an established clade name, while [nP]*Hypotheticus* could indicate that *Hypotheticus* has not been established as a clade name under this code ("n" meaning "not"). If so, the meanings of the symbols [P] and [nP] should be clearly indicated.

Note 21.3A.1. Although Example 1 uses one symbol to indicate establishment under this code and another symbol to indicate the absence of such establishment, an alternative would be to use the presence or absence of a single symbol. However, using absence of a symbol to designate nomenclatural status is potentially confusing because its absence may result from accidental omission. Furthermore, some readers may misinterpret absence of a symbol because they are unaware of the author's convention.

Note 21.3A.2. If a symbol (e.g., quotation marks) is used to indicate non-monophyly of the taxon designated by the genus name, it would be redundant to indicate that the genus name is not an established clade name under this code.

Note 21.3A.3. If a symbol is used to indicate non-monophyly or questionable monophyly of the taxon designated by the genus name, this does not imply that the author does not accept the species. Therefore, the species name should not be interpreted as not validly published under *ICNAFP* Article 36.1.

Recommendation 21.3B. When publishing the name of a new species, selection of a genus name will require consideration of the nomenclatural consequences under both the appropriate rank-based code and this code. In general, a genus name that is also an established clade name (or is simultaneously being established as a clade name) under this code should be selected if possible. (If the names of more than one clade in a nested series of clades satisfy these conditions, any one of the names may be selected.) If this is not possible, an existing genus name may be used, even if the monophyly of the associated taxon under the rank-based code is unknown or doubtful, or a new genus name may be used. If the species to be named cannot be assigned to any taxon with which a genus name has been associated under the appropriate rank-based code, then the only option is to publish a new name to serve as a generic name under the appropriate rank-based code. This name may be simultaneously established as a clade name under this code.

Example 1. If a new species is to be given the binomen *Sorex hockingensis*, and the name *Sorex* has already been established both as a clade name under this code and as the name of a genus under the *ICZN*, then the binomen should appear as *Sorex hockingensis*, new species (or an equivalent expression such as n. sp.), with or without a symbol (e.g., [P]) indicating that *Sorex* is an established clade name (see Rec. 21.3A, Ex. 1).

Example 2. If the taxon associated with the genus name *Sorex* in Example 1 is thought to be monophyletic but has not previously been established as a clade name, the clade name *Sorex* could be established simultaneously with the publication of the binomen *Sorex hockingensis*.

Example 3. If the only preexisting genus to which a new species (for which the specific name or epithet *vulgaris* is selected) can be assigned (*Hypotheticus*) is thought to be non-monophyletic or its monophyly has not been investigated, and the species is part of a clade (*Cladius*) that could be named as a genus under the appropriate rank-based code, then the binomen could appear as *Cladius vulgaris*, new genus and species (or an equivalent expression), with or without a symbol (e.g., [P]) indicating that *Cladius* is an established clade name (see Rec. 21.3A, Ex. 1). If this is done, *Cladius* should be validly published (*ICNP*, *ICNAFP*) or made available (*ICZN*) simultaneously as a genus name under the appropriate rank-based code, and it should also be established as a clade name under this code if it has not previously been established. Alternatively, if it were considered premature to establish the name *Cladius*, the binomen could appear as *Hypotheticus vulgaris*, new species (or an equivalent expression), with or without a symbol (e.g., [nP]) indicating that *Hypotheticus* is not an established clade name (see Rec. 21.3A, Ex. 1) or a symbol (e.g., quotation marks) indicating that *Hypotheticus* is not monophyletic (see Note 21.3A.2).

Example 4. In the situation described in Example 3, if there is not sufficient evidence that the new species is part of any clade that could be named as a genus under the appropriate rank-based code, then the binomen could appear as *Hypotheticus vulgaris*, new species (or an equivalent expression), with or without a symbol (e.g., [nP]) indicating that *Hypotheticus* is not an established clade name (see Rec. 21.3A, Ex. 1) or a symbol (e.g., quotation marks) indicating that *Hypotheticus* is not monophyletic (see Note 21.3A.2). Alternatively, a new genus name could be published in combination with the new specific name or epithet under the rank-based code.

Example 5. If a new species, to be named *campestris*, cannot be assigned to any taxon (whether monophyletic or not) with which a genus name has been associated under the appropriate rank-based code, it would be necessary to publish a new genus name (e.g., *Imaginarius*) in combination with the new specific name or epithet under the rank-based code. If *Imaginarius* is simultaneously established under this code as a clade name, then the binomen should appear as *Imaginarius campestris*, new genus and species (or an equivalent expression), with or without a symbol (e.g., [P]) indicating that *Imaginarius* is an established clade name (see Rec. 21.3A, Ex. 1).

Example 6. If, in the previous example, the name *Imaginarius* is not simultaneously established as a clade name under this code, then the binomen should appear as *Imaginarius campestris*, new genus and species (or an equivalent expression), with or without a symbol (e.g., [nP]) indicating that *Imaginarius* is not an established clade name (see Rec. 21.3A, Ex. 1).

21.4. Subsequent to a species binomen becoming available (*ICZN*) or validly published (*ICNAFP, ICNP*) under the appropriate rank-based code, the second part of the species binomen may be treated as the de facto name of the species under this code, termed a species uninomen. In this context, the species uninomen may be combined with the names of one or more clades the species is part of, in place of or in addition to the genus name (see Rec. 21A).

Recommendation 21.4A. When the genus name is used subsequent to the species binomen becoming available (*ICZN*) or validly published (*ICNAFP, ICNP*), some mechanism should be used to indicate whether the genus name is an established clade name under this code (see Rec. 21.3A, Ex. 1).

Recommendation 21.4B. When the second part of a species binomen is treated as the name of a species subsequent to the species binomen becoming available (*ICZN*) or validly published (*ICNAFP*, *ICNP*), it should be accompanied by one or both of the following: (a) the genus name; (b) the author(s) and year of the publication in which the specific name (*ICZN*) or epithet (*ICNP*, *ICNAFP*) was validly published (*ICNP*, *ICNAFP*) or made available (*ICZN*).

Note 21.4B.1. Under the *ICNAFP*, the author(s) of the binomen is (are) commonly cited, but the year is commonly not cited. In contrast, both the author and year are commonly cited under the *ICZN*. Under this code, if the genus name is not used in combination with the specific name or epithet, both the author and year of the specific name or epithet should be cited. If the genus name is used, citation of the author and year of the specific name or epithet is optional.

Example 1. The species that is referred to as *Vultur gryphus* or *Vultur gryphus* Linnaeus 1758 under the *ICZN* may be referred to under this code as *Vultur gryphus* or *Vultur gryphus* Linnaeus or *gryphus* Linnaeus 1758. Any of these forms of the species name may be associated with additional clade names to indicate hierarchical relationships (see Rec. 21A); for example, *Aves/gryphus* Linnaeus 1758 or *Aves/Vultur gryphus* Linnaeus.

Recommendation 21A. When a species uninomen is combined with more than one genus name and/or clade name, hierarchical relationships among the taxa designated by those names can be indicated in a variety of ways, but the taxa should be listed in order of decreasing inclusiveness from left to right. In addition, symbols such as those in Recommendation 21.3A, Example 1 may be used with any of those names (but for simplicity, such symbols are not included in the following examples).

Example 1. The species originally named *Anolis auratus* Daudin 1802 has been placed in at least two different genera, named *Anolis* and *Norops*. If those names were to be established under this code as the names of (nested) clades, the name and relationships of the species could be indicated in any of the following ways (not an exhaustive list): *Anolis/auratus* Daudin 1802, or *Norops: auratus* Daudin 1802, or *Anolis/Norops/auratus* Daudin 1802, or *Anolis Norops auratus* Daudin 1802. For optional use of parentheses to indicate that a specific name or epithet was originally combined with a different genus name, see Note 21A.3.

Example 2. If the name of a species under the *ICZN* is *Diaulula sandiegensis* (Cooper 1863), and if *Diaulula* has not been established as a clade name under this code (e.g., because there is presently insufficient data to establish monophyly), and if the name *Discodorididae* has been established as the name of a more inclusive clade under this code, then the name and relationships of the species could be indicated in any of the following ways (not an exhaustive list): *Diaulula sandiegensis* Cooper 1863, or *Discodorididae Diaulula sandiegensis* Cooper 1863, or *Discodorididae/sandiegensis* Cooper 1863, or *Discodorididae sandiegensis* Cooper 1863. For optional use of parentheses to indicate that a specific name or epithet was originally combined with a different genus name, see Note 21A.3.

Note 21A.1. When a species uninomen is combined with a clade name that is not also a genus name, the uninomen can be viewed as a name in its own right, rather than as a modifier (adjectival or possessive) of the clade name. Consequently, the ending of the uninomen should not be changed to agree in gender or number (plural vs. singular) with the name with which it is combined. When a species uninomen is combined with a genus name, its gender may be changed to agree with that of the genus name.

Example 1. If the species uninomen *sandiegensis* is combined with the clade name *Discodorididae*, the uninomen should not be changed to *sandiegenses* to match the plural name *Discodorididae*.

Note 21A.2. By combining the second part of a species binomen with the name of a clade that is not a genus under the appropriate rank-based code (see variants that do not use the name *Diaulula* in Rec. 21A, Ex. 2), it is possible to provide phylogenetic information for a species without using a genus name that has not been established as a clade name under this code.

Note 21A.3. If a specific name (*ICZN*) or epithet (*ICNP, ICNAFP*) is associated with just one genus name, parentheses enclosing the name of the author and year of publication of the specific name (*ICZN*) or the author of the epithet (*ICNP, ICNAFP*) may be used to indicate that the specific name or epithet was originally combined with a different genus name, following the conventions of the appropriate rank-based code (which differ in whether the year is commonly cited and whether the author of the accepted combination should also be cited). The use of parentheses for this purpose is optional, which is consistent with the decreased emphasis on categorical ranks under this code. Parentheses may also be used if a specific name or epithet is associated with just one clade name of suprageneric rank under the appropriate rank-based code.

Example 1. In the name *Norops auratus* (Daudin 1802), the use of parentheses indicates that Daudin originally published (made available; *ICZN*) the species name *auratus* in combination with a different genus name. Use of parentheses in this case is optional (e.g., see the citation of this same name and author without parentheses in Rec. 21A, Ex. 1).

Example 2. In the name *Physostegia/virginiana* (Linnaeus) Bentham, the use of parentheses indicates that Linnaeus originally published the specific epithet *virginiana* in combination with a different genus name and that Bentham first validly published (*ICNAFP*) the binomen *Physostegia virginiana*.

Example 3. In the name *Discodorididae sandiegensis* (Cooper 1863) (see Rec. 21A, Ex. 2), the use of parentheses indicates that Cooper originally published the species name *sandiegensis* in combination with a different taxon name.

Chapter XI

Governance

ARTICLE 22. GOVERNANCE OF PHYLOGENETICALLY DEFINED NAMES

22.1. The International Society for Phylogenetic Nomenclature (ISPN) is an international, nonprofit organization with no membership restrictions. Two committees of the ISPN have responsibilities that pertain to this code: the Committee on Phylogenetic Nomenclature (CPN) and the Registration Committee.

22.2. The Registration Committee is responsible for managing the registration database for phylogenetically defined names. It has the authority to set policy concerning the routine operation of the database, so long as such decisions do not conflict with the provisions of this code. The members of the Registration Committee will be appointed by the ISPN through a vote of the Council.

22.3. The CPN has the responsibility and power to: (a) ratify the first edition of this code prior to its implementation; (b) rule on applications for suppression or conservation of names; (c) resolve ambiguities in the provisions of this code; (d) amend the provisions of this code; and (e) produce future editions of this code.

22.4. The members of the CPN will be elected by the membership of the ISPN. The number of members in the CPN will be determined by the ISPN. The CPN officers (Chair and Secretary) will be elected by the membership of the CPN.

22.5. Members of the CPN will be elected for three-year terms. Members may be elected for up to three consecutive terms. Each officer will be elected for a one-year term in that office (as part of the three-year term as a member). Officers may serve for up to three consecutive one-year terms and shall not be eligible to serve again in the same office until one year has elapsed since completing the third consecutive term.

22.6. Applications for suppression or conservation of names, restricted emendations of definitions, and rulings on whether a proposed emendation is restricted or unrestricted must be submitted to the CPN. Once received, they will be published (Art. 4) and made available on a website administered by the ISPN.

22.7. Decisions by the CPN to suppress or conserve names or to emend definitions must be approved by a two-thirds vote of the CPN. Decisions will be published and announced on a website administered by the ISPN, and the affected names will be annotated in the registration database.

22.8. Decisions by the CPN regarding interpretation of rules (in case of ambiguity) and the status of proposed emendations as restricted or unrestricted require approval by a simple majority of the CPN. Decisions will be published and announced on a website administered by the ISPN.

22.9. Proposed modifications of this code after its ratification and implementation must be submitted to the CPN. Once received, they will be published (Art. 4) and made available on a website administered by the ISPN.

22.10. Proposed modifications of this code after its ratification and implementation may not be voted upon until at least three months have elapsed from the date of their publication, to allow for discussion by the systematics community and communication of opinions to the members of the CPN.

22.11. Decisions to modify the code after its ratification and implementation must be approved by a two-thirds vote of the CPN. Any decision adopted by the CPN will be published and announced on a website administered by the ISPN. Decisions take effect immediately upon publication.

Glossary

acceptable name: An established name that is not a (non-conserved) later homonym and thus may potentially be an accepted name.

accepted name: The name that must be adopted for a clade under this code.

ancestor: An entity from which another entity is descended.

apomorphy (apomorphic): A derived character state; a new feature that arose during the course of evolution (see *synapomorphy*).

apomorphy-based definition: A definition that associates a name with a clade originating in the first ancestor to evolve a specified apomorphy that was inherited by one or more internal specifiers. See Article 9.7.

apomorphy-modified crown-clade definition: A minimum-clade definition that is modified by the use of an apomorphy to define the name of a crown clade. See Article 9.9.

apomorphy-modified node-based definition: A term used in earlier versions of this code that is roughly equivalent to an apomorphy-modified crown-clade definition in this version (see footnote to Art. 9.9).

binomen (binomina): A name composed of two words; commonly used to refer to species names composed of a generic name and a specific name (*ICZN*) or epithet (*ICNP, ICNAFP*) under the rank-based codes.

branch: An edge or internode (connection between two nodes) on a tree (in the sense of graph theory). On a phylogenetic tree, a branch is commonly used to represent (1) a lineage between two splitting events (internal branch), or between a splitting event and an extinction event (terminal branch) or between a splitting event and a specified time, often the present (terminal branch), or (2) an ancestor-descendant relationship. The term is sometimes also used (but not in this code) for an internode and all nodes and internodes distal to (descended from) it.

branch-based definition: A term used in earlier versions of this code that is roughly equivalent to a maximum-clade definition in this version (see footnote to Art. 9.5).

branch-modified node-based definition: A term used in earlier versions of this code that is roughly equivalent to a maximum-crown-clade definition in this version (see footnote to Art. 9.5).

categorical rank (also taxonomic rank): A formal category denoting position in a hierarchy of nested taxa. The categorical ranks commonly used in taxonomy comprise seven principal categories (kingdom, phylum or division, class, order, family, genus, and species), which are often treated as mandatory, as well as additional primary categories (e.g., cohort, tribe) and secondary categories (e.g., superorder, subfamily). Note that the species category is treated as a rank in rank-based nomenclature, even when it is conceptualized as a category of biological entities.

clade: An ancestor (an organism, population, or species) and all of its descendants.

conditionally suppressed name: A name that is suppressed only in phylogenetic contexts in which it is a synonym of a particular conserved name (see *suppressed name*).

conserved name: An established name that the Committee on Phylogenetic Nomenclature has ruled should have precedence over earlier synonyms or homonyms.

conversion: The act of establishing a preexisting name in accordance with the rules of this code.

converted (clade) name: A preexisting name that has been established in accordance with the rules of this code (see *new (clade) name*).

crown-based total-clade definition: A total-clade definition that is formulated through reference to the corresponding crown clade. See Article 9.10.

crown clade: A clade originating in the most recent common ancestor of two or more extant species (or organisms). See Article 2.2.

crown-clade definition: A phylogenetic definition that necessarily identifies a crown clade (Art. 2.2) as the referent of a taxon name. See Article 9.9.

definition: A statement specifying the meaning of a name; in this code, it is a statement specifying the taxon to which the name refers.

definitional author: The person(s) who published a phylogenetic definition for a name—either the original definition or an emended one (see *nominal author*).

description: A statement of the features of a taxon (or its component organisms), not limited to those that distinguish it from other taxa with which it might be confused (see *diagnosis*).

diagnosis (diagnoses): A statement of the features of a taxon (or its component organisms) that collectively distinguish it from other taxa with which it might be confused.

directly-specified-ancestor definition: A kind of minimum-clade definition in which the ancestor in which the clade originated is specified directly rather than indirectly through its descendants. See Note 9.5.1.

emendation: A formal change in the phylogenetic definition of a name.

epithet: In the *ICNAFP*, a word that, when combined with the name of a genus, forms the name of an infrageneric taxon (e.g., species, subgenus, section, series) or, when combined with the name of a species, forms the name of an infraspecific taxon (e.g., subspecies, variety, form). The *ICNP* also uses the term "epithet" but only at and below the species rank.

established name: A name that is published in accordance with Article 7 of this code, which may or may not be an acceptable or accepted name.

extant (of a taxon): Having one or more living representatives (organisms) at the present time, or at some specified time since humans began keeping written historical records.

external specifier: A species or specimen that is explicitly excluded from the clade whose name is being defined (see Art. 11.2). Maximum-clade and maximum-crown-clade definitions have external specifiers, but minimum-clade, minimum-crown-clade, apomorphy-based and apomorphy-modified crown-clade definitions normally do not (unless external specifiers are used to prevent use of a name under certain hypotheses of relationships, clade composition, or both; see Art. 11.13).

genus (genera): One of the categorical ranks of rank-based nomenclature; more specifically, it is the primary rank above the rank of species and below that of family. The genus rank is mandatory in rank-based nomenclature not only because it is one of the seven principal ranks (kingdom, phylum or division, class, order, family, genus, species), which are commonly treated as mandatory, but also because the generic name is part of the species binomen.

heterodefinitional: Based on different phylogenetic definitions (see *synonym*).

homodefinitional: Based on the same phylogenetic definition (see *synonym*).

homologous: Shared by virtue of inheritance from a common ancestor. A character or character state shared by two organisms (which may represent different species or clades) is said to be homologous if that character or character state was present in all of their ancestors back to and including their most recent common ancestor.

homonym: A name that is spelled identically to another name that refers to a different taxon. See Article 13.1.

hybrid formula: An expression consisting of the names of two taxa separated by a multiplication sign, designating a single organism or set of organisms of hybrid origin.

ICNAFP: *International Code of Nomenclature for Algae, Fungi, and Plants.*

ICNP: *International Code of Nomenclature of Prokaryotes.*

ICVCN: *International Code of Virus Classification and Nomenclature.*

ICZN: *International Code of Zoological Nomenclature.*

infraspecific epithet: Under the *ICNAFP*, the third word in an infraspecific trinomen.

infraspecific name: Under the *ICZN*, the third word in an infraspecific trinomen.

infraspecific taxon: Under the rank-based codes, a taxon below the rank of species.

internal specifier: A species, specimen, or apomorphy that is explicitly included in the clade whose name is being defined (see Art. 11.2). Every phylogenetic definition has at least one internal specifier, and all of the specifiers in minimum-clade, minimum-crown-clade, apomorphy-based, and apomorphy-modified node-based definitions are normally internal (unless external specifiers are used to prevent use of a name under certain hypotheses of relationships, clade composition, or both; see Art. 11.13).

lineage: A series of entities (e.g., organisms, populations) that form a single unbroken and unbranched sequence of ancestors and descendants. That a lineage is unbranched does not deny the existence of side-branches, which are not parts of the lineage in question, or of branching at lower organizational levels (e.g., organelle lineages within a population lineage). There may even be branching at the organizational level in question as long as it is judged to be temporary.

maximum-clade definition: A definition that associates a name with the largest clade that contains one or more internal specifiers but does not contain one or more external specifiers. See Article 9.6.

maximum-crown-clade definition: A definition that associates a name with the largest crown clade that contains one or more internal specifiers but does not contain one or more external specifiers. See Article 9.9.

maximum-total-clade definition: A definition that associates a name with the largest clade that contains one or more internal specifiers but does not contain one or more external specifiers. See Article 9.10.

minimum-clade definition: A definition that associates a name with the smallest clade that contains two or more internal specifiers. See Article 9.5.

minimum-crown-clade definition: A definition that associates a name with the smallest crown clade that contains two or more internal specifiers. See Article 9.9.

monophyletic: A set consisting of an ancestor and all of its descendants; usually used for groups the members of which share a more recent common ancestor with one another than with any non-members, though monophyletic groups of organisms within sexually reproducing species/populations may not have this property.

name: A word or words used to designate (refer to) an organism or a group of organisms. See *acceptable name, accepted name, conditionally suppressed name, conserved name, converted name, established name, infraspecific name, new name, panclade name, preexisting name, replacement name, scientific name, specific name, suppressed name, taxon name, typified name, unconditionally suppressed name.*

new (clade) name: A newly proposed name that has been established in accordance with the rules of this code (see *converted (clade) name*).

node: A point or vertex on a tree (in the sense of graph theory). On a phylogenetic tree, a node is commonly used to represent (1) the split of one lineage to form two or more lineages (internal node) or the extinction of a lineage (terminal node) or the lineage at a specified time, often the present (terminal node), or (2) a taxon, whether ancestral (internal node) or descendant (internal node or terminal node).

node-based definition: A term used in earlier versions of this code that is roughly equivalent to a minimum-clade definition in this version (see footnote to Art. 9.5).

nomen cladi conversum: See *converted (clade) name*.

nomen cladi novum: See *new (clade) name*.

nomen substitutum: See *replacement name*.

nominal author: The person(s) who first published a name, regardless whether it was phylogenetically defined (see *definitional author*).

orthography: The spelling of a name.

panclade name: A name that is derived from the name of a crown clade by the addition of the prefix *Pan-* and is used to designate the total clade of that crown clade. See Articles 10.3–10.7.

paraphyletic: A set including an ancestor but excluding some or all of its descendants.

phylogenetic: Of or pertaining to the history of ancestry and descent.

phylogenetic definition: A statement specifying a particular clade as the entity to which a name refers.

phylogenetic hypothesis (hypotheses): A proposition about the relationships among biological entities (e.g., species, populations, organisms) in terms of common ancestry.

phylogenetic system (of nomenclature): An integrated set of principles and rules governing the naming of taxa and the application of taxon names that is based on the principle of common descent. This code describes a phylogenetic system of nomenclature.

phylogenetic tree: The diagrammatic representation of phylogeny as a tree in the sense of a minimally connected graph (number of branches = number of nodes minus one).

phylogeny: Evolutionary history; the history of descent with modification, whether in general or a particular part thereof. The term is also sometimes used for a hypothesis of phylogenetic relationships (as in the term *reference phylogeny*).

precedence: The order of preference among established names, used to select the accepted name from among them. In general, precedence is based on the date of establishment, with earlier-established names having precedence over later ones, but later-established names may be conserved over earlier ones.

preexisting name: A scientific name that, prior to its establishment under this code, was either: (a) "legitimate" (*ICNAFP*, *ICNP*), "potentially valid" (*ICZN*), or "valid" (*ICVCN*); or (b) in use but not governed by any code (e.g., zoological names ranked above the family group).

protologue: Everything associated with a name when it was first established (under this code), validly published (*ICNAFP*, *ICNP*), or made available (*ICZN*), for example, description, diagnosis, phylogenetic definition, registration number, designation of type, illustrations, references, synonymy, geographical data, specimen citations, and discussion.

qualifying clause: A part of a phylogenetic definition that specifies conditions under which the defined name cannot be applied.

rank: The position in a hierarchy; in the case of biological nomenclature, the position in a hierarchy of nested taxa.

rank-based codes: The codes that govern the rank-based system of nomenclature—specifically, the *International Code of Nomenclature for Algae, Fungi, and Plants*, the *International Code of Zoological Nomenclature*, the *International Code of*

Nomenclature of Prokaryotes, and the *International Code of Virus Classification and Nomenclature*.

rank-based system (of nomenclature): An integrated set of principles and rules governing the naming of taxa and the application of taxon names that is based on taxonomic ranks (e.g., kingdom, phylum). Also referred to as the "traditional system."

reference phylogeny: A phylogenetic hypothesis that provides a context for applying a clade name by means of its phylogenetic definition.

replacement name: A new name explicitly substituted for a previously established name that is not acceptable because it is a later homonym. A replacement name is equivalent to a *nomen substitutum* in this code. (The term "replacement name" has been used in a broader sense under the *ICZN* to include what the *ICNAFP* refers to as a superfluous name and the *ICZN* refers to as an unnecessary substitute name.)

restricted emendation: A formal change in a phylogenetic definition that requires approval by the Committee on Phylogenetic Nomenclature; a restricted emendation is intended to change the application of a name through a change in the conceptualization of the clade to which it refers. See *unrestricted emendation*.

scientific name: A taxon name that either is formed and governed by one of the codes of biological nomenclature or is of a similar Latinized form (e.g., zoological names ranked above the family group).

sister clade: One member of a pair of clades originating when a single lineage splits into two. Sister clades thus share an exclusive common ancestry and are mutually most closely related to one another in terms of common ancestry.

species: This term is used both for a kind of biological entity (for example, a population lineage segment) and for the lowest primary rank in traditional nomenclature (and thus also for any taxon assigned to that rank). This code does not endorse any species concept nor provide rules for defining species names, but it uses species names governed by the rank-based codes to refer to taxa that are used as specifiers in definitions of clade names. Article 21 provides guidelines for the use of species names governed by the rank-based codes in conjunction with clade names governed by this code.

specific name: Under the *ICZN*, the second word in a species binomen.

specifier: A species, specimen, or apomorphy cited in a phylogenetic definition of a name as a reference point that serves to specify the clade to which the name applies.

stem-based definition: A term used in earlier versions of this code that is roughly equivalent to maximum-clade definition in this version (see footnote to Art. 9.5).

stem-modified node-based definition: A term used in earlier versions of this code that is roughly equivalent to a maximum-crown-clade definition in this version (see footnote to Art. 9.5).

suppressed name: A name that would normally have precedence but does not, due to a decision by the Committee on Phylogenetic Nomenclature to give precedence to a later synonym or homonym.

synapomorphy (synapomorphic): A shared apomorphy. In this code, a synapomorphy is a shared, derived character state inherited from a common ancestor that possessed that state; a shared, independently derived character state is not considered to be a synapomorphy in the sense the term is used in this code (see *apomorphy*).

synonym: A name that is spelled differently than another name that refers to the same taxon. In the case of clade names, synonyms may be homodefinitional or heterodefinitional.

taxon (taxa): A group of organisms or species considered a potential recipient of a scientific name. The only taxa whose names are governed by this code are clades. However, species, whose names are governed by the rank-based codes, are frequently used to define clade names in this code.

taxon name: The word (or, in rank-based codes, words) used to designate a taxon.

taxonomic rank: See *categorical rank.*

total clade: A clade composed of a crown clade and all organisms (and species) that share a more recent common ancestor with that crown clade than with any extant organisms or species that are not members of that crown clade. See Article 2.2.

total-clade definition: A phylogenetic definition that necessarily identifies a total clade (Art. 2.2) as the referent of a taxon name. See Article 9.10.

type (= nomenclatural type): In the rank-based codes, the specimen, specimens, or subordinate taxon to which a taxon name is permanently attached; the type provides the standard of reference that determines the application of a name.

typified name: A name whose application is determined by a type under a rank-based code.

unconditionally suppressed name: A name that has been suppressed by the Committee on Phylogenetic Nomenclature in all phylogenetic contexts (see *suppressed name*); there are no conditions under which it would have precedence over any other name.

uninomen (uninomina): A name composed of a single word; in this code, the term is used to refer to the second part of a species binomen that is being treated as the name of a species (though the names of clades are also uninomina).

unrestricted emendation: A formal change in a phylogenetic definition that does not require approval by the Committee on Phylogenetic Nomenclature; an unrestricted emendation is intended to prevent undesirable changes in the application of a particular name (in terms of clade conceptualization) when the original definition is applied in the context of a revised phylogeny. See *restricted emendation.*

Appendix A

Registration Procedures and Data Requirements

MOST RECENT REVISION: JANUARY 6, 2019

This appendix may be revised more frequently than the main body of the code and without a formal meeting of the CPN. The most recent information will be made available through the *PhyloCode* and/or ISPN website.

I. Registration procedures

Clade names and definitions governed by this code must be registered in RegNum, the repository of clade names that is maintained under the auspices of the International Society for Phylogenetic Nomenclature. After a name and definition are submitted to RegNum through its web interface, the submission will be checked for missing data. A temporary registration number will be issued at this time if the work in which the name will appear has not yet been accepted. Once the work has been accepted for publication, the author must notify RegNum and provide a full reference to the publication in order for the registration number to become permanent. Alternatively, an author may wait until after acceptance for publication before submitting the name, in which case the registration number will be issued immediately.

If the spelling or definition of a submitted name is identical to one that already exists in the registration database, the author will be warned.

II. Data fields (mandatory data indicated with an asterisk)

1. Data common to all clade names:

Contact information (for each author): name*, mailing address, phone number, fax number, email address*, home page URL.

Name to be registered*

Type of name* (new clade name, converted clade name)

Date of registration

Bibliographic reference to publication

Date of publication

Definition type* (minimum-clade, maximum-clade, apomorphy-based, etc.)

Phylogenetic definition*

List of specifiers*

■ For a species cited as a specifier: name*, author*, year of publication*, code which governs the name*, URL of taxonomic database holding information

■ For an apomorphy cited as a specifier: description*

■ For a type specimen cited as a specifier: species name typified*, author of species name typified*, year of publication of species name typified*, code governing typified name*

■ For a specimen (other than a type) cited as a specifier: repository institution*, collection data needed to locate the specimen*, description or image or bibliographic reference to published image*

Qualifying clause (required if part of definition)

Status of definition as emended (if appropriate)

Reference phylogeny (bibliographic reference, URL, or accession number in public repository)*

Status of name as conserved or suppressed (if appropriate)

Author's comments

Administrator's annotations

2. Data particular to converted clade names:

Preexisting name*

Author of preexisting name*

Direct bibliographic reference to original publication of preexisting name (including year)*

Code governing the preexisting name*

URL of taxonomic database holding information about the name

3. Data particular to new clade names:

For a replacement name: replaced name*

Appendix B

Code of Ethics

1. Authors proposing new names or converting preexisting names should observe the following principles, which together constitute a code of ethics.
2. Authors should not publish a new name or convert a preexisting one if they have reason to believe that another person intends to establish a name for the same clade (or that the clade is to be named in a posthumous work). An author in such a position should communicate with the other person (or their representatives) and only attempt to establish a name if the other person has failed to do so in a reasonable period (not less than a year).
3. Authors should not publish a replacement name (a *nomen substitutum*) for a later homonym without informing the author of the latter name about the homonymy and allowing that person a reasonable time (at least a year) to establish a replacement name.
4. Authors should not propose a name that, to their knowledge or reasonable belief, would be likely to give offense on any grounds.
5. Authors should not use offensive or insulting language in any discussion or writing that involves phylogenetic nomenclature. Debates about phylogenetic nomenclature should be conducted in a courteous and professional manner.

6. Editors and others responsible for the publication of works dealing with phylogenetic nomenclature should avoid publishing any material that appears to them to contain a breach of the above principles.

7. Adherence to these principles is a matter for the conscience of individual persons. The CPN is not empowered to rule on alleged breaches of them.

Appendix C

Equivalence of Nomenclatural Terms among Codes

Equivalence table of nomenclatural terms used in this code, the Draft *BioCode* and the current biological codes, except the *International Code of Virus Classification and Nomenclature* (patterned after a similar table in the Draft *BioCode*). The criteria represented by terms treated here as equivalent are not always exactly the same (e.g., establishment of a clade name in this code requires a phylogenetic definition, which is not a requirement of any other code). *BioCode* = Draft *BioCode* (Taxon 60: 201–212 [2011]). *Prokaryotic Code = International Code of Nomenclature of Prokaryotes: Prokaryotic Code (2008 Revision)* (published 2019). *Botanical Code = International Code of Nomenclature for Algae, Fungi, and Plants (Shenzhen Code)* (2018). *Zoological Code = International Code of Zoological Nomenclature* (1999, but including subsequent amendments that took effect in 2012).

This Code	BioCode	Prokaryotic Code	Botanical Code	Zoological Code
Publication and precedence of names				
published	----------	effectively published	effectively published	published
precedence	precedence	priority	priority	precedence
earlier	earlier	senior	earlier	senior
later	later	junior	later	junior
Nomenclatural status				
established	established	validly published	validly published	available
converted	----------	----------	----------	----------
acceptable	acceptable	legitimate	legitimate	potentially valid
registration	registration	validation	registration	registration
Taxonomic status				
accepted	accepted	correct	correct	valid
Synonymy and homonymy				
homodefinitional	----------	objective	nomenclatural	objective
heterodefinitional	----------	subjective	taxonomic	subjective
replacement name	replacement name	deliberate substitute	avowed substitute	new replacement name
----------	----------	----------	superfluous name	unnecessary substitute name
Conservation and suppression				
conserved	conserved	conserved	conserved	conserved
suppressed	suppressed	rejected	rejected	suppressed

Index

The references are not to pages but to the Articles, Recommendations, etc. of this Code, as follows: Pre = Preamble; Pri = Principles; Numerals = Articles; Numerals followed by letters = Recommendations; Ex = Examples; N = Notes; App = Appendix; G = Glossary. The Preface is not covered by the Index.

If a term occurs in both an article, recommendation, or note and an accompanying example, often only the article, recommendation or note is cited. However, if the term appears in an example but not the article, recommendation, or note with which it is associated, the example is cited (e.g., 9.11.Ex1).

form of, 17.1, 17.3, 17.3A,
17.3B, 17.4, 17B
governed by this code, 2.2
independent of categorical
ranks, 3.1
low-level (coinciding with or
nesting within species),
N21.1.1
may not be converted from
preexisting specific or
infraspecific epithet, 10.10
new, 8.4, 8C, 9.1, 9.2, 10.2,
10.4A, 17B, 18.4, G, AppA
nominal and definitional
authors the same,
N19.1.2, N20.2.1
selection of, 10A, 10B, 10C,
N11G.2, 17.3B
not to be based on names of
ichnotaxa or ootaxa, 11C
prefixes, 10.3, 10.8, 10D, 11G,
N11G.1, N11G.2, 17.1
registered but not published,
8B, 8C
selection of, 10, 10.1, 10.1A,
10A, 10B, 10C, 11C, 12
if based on genus name
under rank-based codes,
11.10A, 11.10B
to minimize disruption
of current/historical
usage, 10.1
source of (preexisting and
new), 9.1
suffixes, 11G, N11G.1
total, 10.2, 10.3, 10.3A,
10.6, 10.7
unaffected by connotations, 6.6
connotation of, 6.6, 10C
conserved, 12.2, 15.3, 15.6, 15.7, G,
AppC (See also Conservation.)

converted, 6.3, N7.2.2, 8.4, 9.1,
9.2, 9.15, N9.15.1, N9.15.2,
9.15A, N.15A.3, N9.15A.4,
10.4A, 10.6, 10.9, 10.10, 10D,
10E, 10F, 10G, 11.10. 11.10B,
N11.11.2, 11A, N11A.1,
N11G.2, 19.1, N19.1.2, 20.2,
20.4, N20.4.1, 20.4B, G, AppA,
AppC (See also Conversion.)
spelling, 17.5, 17.5A, N18.1.1
nominal and definitional
authors often different,
N19.1.2
earlier (established), 12.2, N12.2.1,
N13.1.1, 14.5, 15.4, AppC
established, 5.1, 6.1, 6.1B, 6.3, 6.4,
6.5, 6.6, 7.2, N7.2.1, N7.2.2,
8.1, 9.1, 9.2, 9.3, 9.13, N9.13.2,
9.14, 9.15, N9.15.2, 9A, 10.1A,
10.2, 10.3, 10D, 10F, 12.1,
12.2, N12.2.1, 13.1, N13.1.1,
13.5, 13.6, 14.1, 14.4, 14.5,
15.5, 15.6, 15.7, 17.1, 17.4, 17.5,
17A, 18.1, N18.1.2, 18.6, 20A,
N21.1.1, 21.3A, N21.3A.1,
21.3B, 21.4A, G, AppC
generic (or genus), 10.10.Ex1,
10D.Ex1, 10F, 10G,
11.10, N11G.2, 21.2, 21.3,
N21.3A.2–3, 21.4, 21.4A,
21.4B, N21.4B.1, 21A,
21A.Ex1–2, N21A.1, N21A.2,
N21A.3, N21A.3.Ex1–2, G
(under "genus")
indicating whether it is an
established clade name,
21.3A, N21.3A.2
selection of, when publishing a
new species name, 21.3B
governance by different codes, 6.1B
inaccurate connotations, 6.6